INTRODUCTION TO METAPHYSICS

Introduction to METAPHYSICS

Lauren Bell
Rick Ponting
Michael Warne

Kruger Brentt
Publishers
2023

Kruger Brentt Publishers UK. LTD.
Company Number 9728962

Regd. Office: 68 St Margarets Road, Edgware, Middlesex HA8 9UU

ISBN: 9781787150812

For information on all our publications visit our website at http://krugerbrentt.com/

PREFACE

Philosophy is an academic field concerned with the generation of a clear understanding of such phenomenon as nature, the confines of knowledge, and the rules of moral judgment. As such, the field is made up of five branches: metaphysics, epistemology, ethics, logic, and aesthetics, all of which contribute to the achievement of its goals. Each of the five branches of philosophy is founded on unique but interrelated phenomena. This paper will concentrate on metaphysics as a branch of philosophy and will seek to give a vivid description of the sub- branch as well as its importance. The indispensability of metaphysics is due to its primary objective of explaining the nature of reality. Metaphysics forms the foundation upon which all scientific ideas are resolved and their validity as well as distinct elements established. The fact that all scientific endeavors are aimed at explaining a particular phenomenon, metaphysics comes into play by evoking the inquisitive nature of research with the "what is?" question. It, therefore, forms the core driving force for any scientific inquiry. Metaphysical studies generally seek to explain inherent or universal elements of reality which are not easily discovered or experienced in our everyday life. As such, it is concerned with explaining the features of reality that exist beyond the physical world and our immediate senses.

Metaphysic is divided into that of the speculative and that of the practical use of pure reason, and is, accordingly, either the metaphysic of nature, or the metaphysic of ethics. The former contains all the pure rational principles—based upon conceptions alone (and thus excluding mathematics)—of all theoretical cognition; the latter, the principles which determine and necessitate a priori all action. Now moral philosophy alone contains a code of laws—for the regulation of our actions—which are deduced from principles entirely a priori. Hence the metaphysic of ethics is the only pure moral philosophy, as it is not based upon anthropological or other empirical considerations. The metaphysic of speculative reason is what is commonly called metaphysic in the more limited sense.

The present book contains eleven chapters covering all related disciplines. It is written in a lucid style with theoretical discussions being supplemented with illustrations, and tables for easy understanding of the subject. It covers all of the most important topics viz. major doctrines, Platonism, Aristotle's metaphysics, Thomism, Cartesians, Vedic metaphysics,Buddhist metaphysics, metaphysics in Chinese philosophy, Janina philosophy and metaphysics and metaphysical problems and its relation to education. The book gives a summary of the metaphysics and its applications. The book is useful to teachers, students and useful to all those who are interested in metaphysics.

We feel extremely grateful to all our colleagues for their comments and suggestions. Finally, we would like to thank everyone who encouraged us in every step of the way in the successful completion of the book. We express indebtedness to authors of books, manuals, periodicals, magazines, newsletters, journals and websites etc. that helped in the preparation of this book. Healthy criticism and suggestions for further improvement of the book are solicited.

Lauren Bell

Rick Ponting

Michael Warne

CONTENTS

1

METAPHYSICS: AN INTRODUCTION

1.1 INTRODUCTION

The philosopher Immanuel Kant once referred to Metaphysics as "a bottomless abyss" and "a dark ocean without a shore" while the American philosopher William James called it "nothing but an unusually obstinate way to think clearly."

In this chapter we are going to provide an introduction to Metaphysics. To do so we will look at the history of the discipline, examine its subject matter, and discuss the 20th century philosopher Ludwig Wittgenstein's influential views on Metaphysics.

Metaphysics is a branch of philosophy whose roots stretch back to the time of the ancient Greeks approximately 2500 years ago. Along with ethics, epistemology, and logic, metaphysics is considered one of the main branches of philosophy. While there is no agreed-upon definition of metaphysics, it is often described as the discipline which is concerned with the ultimate nature of reality. However, to those not familiar with metaphysics, such a definition will prove rather vague. So in order to better grasp its subject matter it will be helpful to examine the history of the discipline.

A few hundred years following his death, around the first century BC, Aristotle's works were being published in the ancient city of Alexandria. One of the works published was given the title 'Metaphysics', which literally means 'after the physics'. This title was not meant to reflect the content of the treatise, as Aristotle had referred to the subject matter of it as first philosophy or theology. Rather it was given this title because it had followed, or come after, the publication of a treatise by Aristotle titled "Physics". However, due to Aristotle's immense influence the word metaphysics soon evolved from merely being the title of one of his books, to becoming the name of an entire philosophical discipline.

While Aristotle's work covered a wide array of topics there are two underlying themes to it which many have pointed to as constituting the essence of metaphysics. The first of these themes is the study of first causes, or in other words, that which does not change and from which emanates the things in this world we experience. The interest in first causes is why God has been a prominent topic of metaphysics throughout history. The second theme is the study of being or existence, in which an attempt is made to identify and delineate the fundamental categories of being. The study of being is simply the study of that which is, or that which exists.

While Aristotle's conception of metaphysics continues to be influential up to the present day, the discipline did see some important developments in the 17th and 18th centuries. During this time a group of philosophers – known as the continental rationalists – began to distinguish between what they called general metaphysics and special metaphysics.

General metaphysics, also referred to as ontology, is the study of being or existence and is in line with Aristotle's conception of metaphysics. Special metaphysics, on the other hand was divided into three disciplines; cosmology, rational psychology, and natural theology. While general metaphysics was concerned with being at a broad, fundamental level, special metaphysics addressed more specific questions concerning existence. Topics addressed within special metaphysics included such things as immortality, freedom of the will, and the mind body problem.

The interest with these more specific issues continues to the present day. However, the problems of general metaphysics, or what is more commonly called ontology these days, remains the most prominent area of metaphysics. Present-day ontologists, it should be noted, are not generally concerned with what the concept of existence means. Rather, like Aristotle, they try to provide accounts or inventories of the things that do exist and in the process identifying the fundamental categories of being. Thus ontologists will try to determine whether such things as properties, numbers, events, relations, souls, material objects, or universals exist; and if they do what the characteristics of such things are.

The question of whether universals exist is one of the oldest questions of ontology and has been debated since the time of the ancient Greeks. Universals are things which can be instantiated or shared by different individual objects. Examples of universals include redness, squareness, and beauty. Obviously there are many different things can be red, square, or beautiful, but the question is whether the universals themselves have any sort of existence apart from the particular things they occupy.

Along with these questions of ontology, metaphysicians today, like the continental rationalists also examine more specific questions. Topics addressed by contemporary metaphysicians include the nature of time and space, the mind-body problem, causality, what it means to be a person, and the problem of free will, amongst others.

The philosopher Peter van Inwagen in his book titled Metaphysics has put forth three questions which he believes encapsulates the subject matter of contemporary metaphysics:

1. What are the most general features of the World, and what sorts of things does it contain? What is the World like?
2. Why does a World exist – and, more specifically, why is there a World having the features and the content described in the answer to question 1?
3. What is our place in the World? How do we human beings fit into it?

As should now be evident metaphysics deals with some very abstract questions, questions which many philosophers, metaphysicians amongst them, believe may not even be answerable. In fact, the 19th century philosopher FH Bradley wrote somewhat humorously: "Metaphysics is the finding of bad reasons for what we believe on instinct..."

Ludwig Wittgenstein, one of the most influential philosophers of all time, was also skeptical about the ability to answer metaphysical questions. As he wrote:

"Most of the propositions and questions to be found in philosophical works are not false but nonsensical. Consequently, we cannot give any answer to questions of this kind, but can only establish that they are nonsensical... (They belong to the same class as the question whether the good is more or less identical than the beautiful)."

What did Wittgenstein, who it should be noted is by no means the clearest of philosophers, mean when he said that most philosophical propositions and questions are not false, but rather non-sensical? KT Fann provides an explanation in his excellent work "Wittgenstein's Conception of Philosophy":

"His point is simply this: Philosophical propositions are not false, they do not misstate facts which could be correctly stated, for they do not state or misstate any facts at all – they merely look like propositions but are in reality, not propositions in the strict sense. The attempt to say something (in the sense of stating propositions) about what transcends the world (the inexpressible) results in nonsense." (KT Fann Wittgenstein's Conception of Philosophy)

In other words, Wittgenstein did not believe that the questions posed by metaphysicians could be answered with the use of language. Instead, the problems of metaphysics according to him "transcend" the world, or as Wittgenstein put it, "The solution of the riddle of life in space and time lies outside space and time."... or alternatively "There are, indeed, things that are inexpressible. They show themselves. They are what is mystical".

However, it should be noted that even though Wittgenstein did not believe that most of the problems of metaphysics could be answered, he was not completely against the discipline. In fact, he was heavily influenced by the 19th century German philosopher Arthur Schopenhauer, who put forth one of the most elaborate and extensive metaphysical systems in the history of philosophy. And to conclude this article we will provide a quote by Wittgenstein which reveals his appreciation of metaphysics:

"Don't think I despise metaphysics or ridicule it. On the contrary, I regard the great metaphysical writings of the past as among the noblest productions of the human mind."

1.2 METAPHYSICS

Metaphysics is the branch of philosophy that studies the fundamental nature of reality, the first principles of being, identity and change, space and time, causality, necessity, and possibility. It includes questions about the nature of consciousness and the relationship between mind and matter, between substance and attribute, and between potentiality and actuality. The word "metaphysics" comes from two Greek words that, together, literally mean "after or behind or among the natural". It has been suggested that the term might have been coined by a first century CE editor who assembled various small selections of Aristotle's works into the treatise we now know by the name Metaphysics.

Metaphysics sounds as though it should be something like physics, only meta. Actually, 'meta' in ancient Greek meant 'after'. The word 'metaphysics' was coined by an ancient editor of Aristotle's works, who simply used it for the books listed after those on physics. The physics books discussed things that change; the metaphysics books discussed things that don't change.

Today, the word 'metaphysics' is used more widely, for the branch of philosophy that studies, in a very general way, what there is and how it is. Thus the idea that everything changes counts as metaphysical, even though it rules out the subject matter of Aristotle's Metaphysics.

There is a deeper problem about how metaphysics relates to physics and the rest of natural science, apart from the origins of the words. For doesn't natural science find out what there is in the world, and how it is structured? If so, what room is left for metaphysics? Natural scientists and metaphysicians may seem to be asking the same questions. The difference is that the natural scientists base their answers on observation, experiment, measurement and calculation, while the metaphysicians base theirs on armchair reflection. So is metaphysics just lazy physics, long past its sell-by date?

Metaphysicians might claim that natural science investigates the world of experience, while metaphysical questions concern aspects of reality that transcend experience. The danger is that any attempt to go beyond our experience ends up meaningless, since we can make no sense of it. Such suspicions of metaphysics have been around since David Hume and Immanuel Kant in the 18th century. But they are not decisive.

Without understanding the metaphysics of numbers, we cannot properly understand the role of mathematics in science. We can define 'naturalism' as the view that everything is natural, the kind of thing studied by natural science, like electrons and genes. Naturalism is itself a meaningful metaphysical theory. But it is not the kind of theory that the usual methods of natural science are designed to assess. For example, are numbers natural? Natural science uses mathematics all the time, but it doesn't study numbers themselves – it just takes them for granted and applies them to measure physical quantities. So are numbers counterexamples to naturalism?

Another view is that there are really no such things as numbers; they are just useful fictions. But explaining how a mere fiction can be as useful as numbers are to natural science has proved a formidably difficult task. Without understanding the metaphysics of numbers, we cannot properly understand the role of mathematics in science.

Metaphysics lurks in unexpected places. When Margaret Thatcher was Prime Minister, she made a notorious metaphysical claim: "There is no such thing as society" . On her view, there are the individual citizens, who are related to each other, but they do not constitute some further entity, a society. Of course, the opposite is also a metaphysical claim: that there is such a thing as society. Both claims concern what kinds of thing there are. It is a typical metaphysical dispute, between reductionist and anti-reductionist views. Which side is right? It might seem to make no practical difference, except that one side was used to justify controversial social and political policies.

A more hard-line reductionist than Mrs Thatcher would say "There is no such thing as the individual'". On that view, there are the particular atoms, which are related to each other, but they do not constitute some further entity, a person – for instance, you. Have you ever wondered what kind of thing you are: a human animal, an immortal soul, a stream of consciousness, nothing? In asking such questions, you are already doing metaphysics.

Metaphysics studies questions related to what it is for something to exist and what types of existence there are. Metaphysics seeks to answer, in an abstract and fully general manner, the questions:

1. What is there?
2. What is it like?

Topics of metaphysical investigation include existence, objects and their properties, space and time, cause and effect, and possibility. Metaphysics is considered one of the four main branches of philosophy, along with epistemology, logic, and ethics.

1.2.1 Existence and Consciousness

Existence (the fact or state of continued being) is axiomatic (meaning that it does not rest upon anything in order to be valid, and it cannot be proven by any "more basic" premises) because it is necessary for all knowledge and it cannot be denied without conceding its truth (a denial of something is only possible if existence exists). "Existence exists" is therefore an axiom which states that there is something, as opposed to nothing.

Consciousness is the faculty which perceives and identifies things that exist. In his famous formulation "Cogito ergo sum" ("I think therefore I am"), René Descartes argued that consciousness is axiomatic, because you cannot logically deny your mind's existence at the same time as using your mind to do the denying.

However, what Descartes did not make clear is that consciousness is the faculty that perceives that which exists, so it requires something outside of itself in order to function: it requires, and is dependent upon, existence. The primacy of existence states that existence is primary and consciousness is secondary, because there can be no consciousness without something existing to perceive. Existence is independent of, makes possible, and is a prerequisite of consciousness. Consciousness is not responsible for creating reality: it is completely dependent upon reality.

1.2.2 Mind and Matter

Early debates on the nature of matter centered on identifying a single underlying principle: water was claimed by Thales, air by Anaximenes, Apeiron (meaning "the undefined infinite") by Anaximander, and fire by Heraclitus. Democritus conceived an atomic theory (Atomism) many centuries before it was accepted by modern science.

The nature of the mind and its relation to the body has also exercised the best brains for millennia. There is a large overlap here with Philosophy of Mind, which is the branch of philosophy that studies the nature of the mind, mental events, mental functions, mental properties and consciousness, and their relationship to the physical body.

In the 17th Century, Descartes proposed a Dualist solution called Substance Dualism (or Cartesian Dualism) whereby the mind and body are totally separate and different: the mental does not have extension in space, and the material cannot think.

Idealists, like Bishop George Berkeley and the German Idealist school, claim that material objects do not exist unless perceived (Idealism is essentially a Monist, rather than Dualist, theory in that there is a single universal substance or principle).

Baruch Spinoza and Bertrand Russell both adopted, in different ways, a dual-aspect theory called Neutral Monism, which claims that existence consists of a single substance which in itself is neither mental nor physical, but is capable of mental and physical aspects or attributes.

In the last century, science (particularly atomic theory, evolution, computer technology and neuroscience) has demonstrated many ways in which mind and brain interact in a physical way, but the exact nature of the relationship is still open to debate. The dominant metaphysics in the 20th Century has therefore been various versions of Physicalism (or Materialism), a Monist solution which explains matter and mind as mere aspects of each other, or derivatives of a neutral substance.

1.2.3 Objects and Their Properties

The world contains many individual things (objects or particulars), both physical and abstract, and what these things have in common with each other are called universals or properties. Metaphysicians are interested in the nature of objects and their properties, and the relationship between the two.

The problem of universals arises when people start to consider in what sense it is possible for a property to exist in more than one place at the same time (e.g. a red car and a red rose). It seems clear that there are many red things, for example, but is there an existing property of 'redness'? And if there is such a thing as 'redness', what kind of thing is it? See the section on Realism for a further discussion of this.

Any object or entity is the sum of its parts. The identity of an entity composed of other entities can be explained by reference to the identity of the building blocks, and how they are interacting. A house can be explained by reference to the wood, metal, and glass that are combined in that particular way to form the house; or it could be explained in terms of the atoms that form it.

1.2.4 Identity and Change

Identity is whatever makes an entity definable and recognizable, in terms of possessing a set of qualities or characteristics that distinguish it from entities of a different type (effectively, whatever makes something the same or different). Thus, according to Leibniz, if some object x is identical to some object y, then any property that x has, y will have as well, and vice versa (otherwise, by definition, they would not be identical).

Aristotle's Law of Identity (or the Axiom of Identity) states that to exist, an existent (i.e. an entity that exists) must have a particular identity. A thing cannot exist without existing as something, otherwise it would be nothing and it would not exist. Also, to have an identity means to have a single identity: an object cannot have two identities at the same time or in the same respect. The concept of identity is important because it makes explicit that reality has a definite nature, which makes it knowable and, since it exists in a particular way, it has no contradictions (when two ideas each make the other impossible).

Change is the alteration of identities, whether it be a stone falling to earth or a log burning to ash. For something to change (which is an effect), it needs to be acted on (caused) by a previous action. Causality is the law that states that each cause has a specific effect, and that this effect is dependent on the initial identities of the agents involved.

We are intuitively aware of change occurring over time (e.g. a tree loses a leaf). The Ancient Greeks took some extreme positions on the nature of change: Parmenides denied that change occurs at all, while Heraclitus thought change was ubiquitous.

Currently there are three main theories which deal with the problem of change:

- Mereological Essentialism assumes that an object's parts are essential to it, and therefore that an object cannot persist through any change of its parts.
- Perdurantism holds that objects are effectively 4-dimensional entities made up of a series of temporal parts like the frames of a movie (it treats the tree, then, as a series of tree-stages).
- Endurantism, on the other hand, holds that a whole object - and the same object - exists at each moment of its history, (so that the same tree persists regardless of how many leaves it loses).

1.2.5 Space and Time

A traditional Realist position is that time and space have existence independent from the human mind. Idealists, however, claim that space and time are mental constructs used to organize perceptions, or are otherwise unreal.

Descartes and Leibniz believed that, without physical objects, "space" would be meaningless because space is the framework upon which we understand how physical objects are related to each other. Sir Isaac Newton, on the other hand, argued for an absolute space ("container space"), which can continue to exist in the absence of matter. With the work of Sir Albert Einstein, the pendulum swung back to relational space in which space is composed of relations between objects, with the implication that it cannot exist in the absence of matter.

Although Parmenides denied the flow of time completely in ancient times, echoed more recently by the British Idealist J.M.E. McTaggart (1866 - 1925), much debate in both philosophy and physics has centered on the direction of time ("time's arrow"), and whether it is reversible or symmetrical. As for whether objects persist over time, then the endurantism / perdurantism dichotomy described above applies.

1.2.6 Religion and Spirituality

Theology is the study of God and the nature of the Divine. This is sometimes considered a whole separate branch of philosophy, the Philosophy of Religion (see that section for more detail). It asks questions like:

- Does the Divine intervene directly in the world (Theism), or is its sole function to be the first cause of the universe (Deism)?
- Is there one God (Monotheism), many gods (Polytheism) or no gods (Atheism or Humanism), or is it impossible to know (Agnosticism)?
- Are God and the universe identical (Pantheism, Monism) or are they different (Panentheism, Dualism)?
- Does religious belief depend on faith and revelation (Fideism), or on reason (Deism)?

Within Western Philosophy, Philosophy of Religion, and theology in general, reached its peak with Medieval Christian schools of thought like Scholasticism.

1.2.7 Necessity and Possibility

A necessary fact is true across all possible worlds (that is, we could not imagine it to be otherwise). A possible fact is one that is true in some possible world, even if not in the actual world. This idea of possible worlds was first introduced by Gottfried Leibniz, although others have dealt with it in much more detail since, notably the American analytic philosopher David Lewis (1941 - 2001) in his theory of Modal Realism.

The concept of necessity and contingency (another term used in philosophy to describe the possibility of something happening or not happening) is also central to some of the arguments used to justify the existence or non-existence of God, notably the Cosmological Argument from Contingency (see the section on Philosophy of Religion for more details).

1.3 NATURE AND SCOPE OF METAPHYSICS

In the 4th century BCE the Greek philosopher Aristotle wrote a treatise about what he variously called "first philosophy," "first science," "wisdom," and "theology." In the 1st century BCE, an editor of his works gave that treatise the title Ta meta ta physika, which means, roughly, "the ones after the ones about nature." "The ones about nature" are those books that make up what is today called Aristotle's Physics, as well other writings of his on the natural world. The Physics is not about the quantitative science now called physics; instead, it concerns philosophical problems about sensible and mutable (i.e., physical) objects. The title Ta meta ta physika probably conveyed the editor's opinion that students of Aristotle's philosophy should begin their study of first philosophy only after they had mastered the Physics. The Latin singular noun metaphysica was derived from the Greek title and used both as the title of Aristotle's treatise and as the name of its subject matter. Accordingly, metaphysica is the root of the words for metaphysics in almost all western European languages (e.g., metaphysics, la métaphysique, die Metaphysik).

Aristotle provided two definitions of first philosophy: the study of "being as such" (i.e., the nature of being, or what it is for a thing to be or to exist) and the study of "the first causes of things" (i.e., their original or primary causes). The relation between these two definitions is a much-debated question. Whatever its answer may be, however, it is clear that the subject matter of what is today called metaphysics cannot be identified with that of Aristotle's Metaphysics. While it is certainly true that all the problems that Aristotle considered in his treatise are still said to belong to metaphysics, since at least the 17th century the word metaphysics has been applied to a much wider range of questions. Indeed, if Aristotle were somehow able to examine a present-day textbook on metaphysics, he would classify much of its content not as metaphysics but as physics, as he understood the latter term. To take only one example, the modern book would almost certainly contain a great deal of discussion of philosophical problems regarding the identity of material objects (i.e., the conditions under which material objects are numerically the same as, or different from, each other; see below Problems in metaphysics: Identity). An ancient example of such a problem is the following: A statue is formed by pouring molten gold into a certain mold. The statue is then melted down and the molten gold poured into the same mold and allowed to cool and solidify. Is the resulting statue the same statue as the original? Such problems evidently do not concern either being as such or the first causes of things.

The question of why modern metaphysics is a much broader field than the one conceived by Aristotle is not easy to answer. Some partial or contributing causes, however, may be the following.

- The appropriation of the word physics by the quantitative science that now bears that name, with the result that some problems that Aristotle would have regarded as belonging to "physics" could no longer be so classified. As regards the problem of the gold statue, for example, modern physics can explain why the melting point of gold is lower than the melting point of iron, but it has nothing to say about the identity of recast statues.
- Similarity of method between Aristotelian and modern metaphysics. The American philosopher William James (1842–1910) said, "Metaphysics means only an unusually obstinate attempt to think clearly and consistently." That is not a bad statement of the only method that is available to students of metaphysics in either its original Aristotelian sense or in its more recent extended sense. If one is interested in questions about the nature of being, the first causes of things, the identity of physical objects, or the nature of causation (the last two problems belong to metaphysics in its modern sense but not its original sense), one will find that the only method available is an "obstinate attempt to think clearly and consistently" about them.
- Overlap of subject matter between Aristotelian metaphysics and Aristotelian physics. The topics "being as such" and "the first causes of things" cannot be wholly divorced from philosophical problems about sensible and mutable objects, the original subject matter of Aristotle's physics. Sensible and mutable objects, after all, are—that is, they exist—and, if indeed there are first causes of things, they certainly stand in causal relations to those first causes.

Whatever the reasons may be, the set of problems to which the word metaphysics now applies is so diverse that it is very hard to frame a definition that adequately expresses the nature and scope of the discipline. Such traditional definitions as "an investigation into the nature of being," "an attempt to describe the reality that lies behind all appearances," and "an investigation into the first principles of things" are not only vague and barely informative but also positively inaccurate: each of them is either too broad (it can be applied just as plausibly to philosophical disciplines other than metaphysics) or too narrow (it cannot be applied to some problems that are paradigmatically metaphysical). Thus, the only way to give a useful account of the nature and scope of metaphysics as the term is now understood is to provide a survey of a series of philosophical problems that uncontroversially belong to modern metaphysics.

1.4 ABSTRACT, OBJECTS AND MATHEMATICS

Some philosophers hold that there are abstract objects (such as numbers, mathematical objects and fictional entities) and universals (properties that can be possessed by multiple objects, such as "redness" or "squareness"), both of which are outside of space and time and/or are causally inert.

Realism, best exemplified by Plato and his Platonic Forms, teaches that universals really exist, independently and somehow prior to the world.

On the other hand, Nominalism holds that there is really no such thing as abstract objects, which really exist only as names, because a single object cannot exist in multiple places simultaneously.

Moderate Realism, as espoused by Aristotle among others, tries to find some middle ground between Nominalism and Realism, and holds that there is no realm as such in which universals exist, but rather they are located in space and time wherever they happen to be manifest. Conceptualism, the doctrine that universals exist only within the mind and have no external or substantial reality, is also an intermediate solution.

Other positions such as Formalism and Fictionalism do not attribute any existence to mathematical entities, and are anti-Realist.

The Philosophy of Mathematics overlaps with metaphysics in this area.

1.4.1 Determinism and Free Will

Determinism is the philosophical proposition that every event, including human cognition, decision and action, is causally determined by an unbroken chain of prior occurrences. Thus, there is at any instant only one physically possible future, and no random, spontaneous, mysterious or miraculous events ever occur.

This posits that there is no such thing as Free Will, where rational agents can exercise control over their own actions and decisions. Incompatibilists (or Hard Determinists) like Baruch Spinoza, view determinism and free will as mutually exclusive. Others, labeled Compatibilists (or Soft Determinists), like Thomas Hobbes, believe that the two ideas can be coherently reconciled.

It should be noted that Determinism does not necessarily mean that humanity or individual humans have no influence on the future (that is known as Fatalism), just that the level to which human beings have influence over their future is itself dependent on present and past.

1.4.2 Cosmology and Cosmogony

Cosmology is the branch of metaphysics that deals with the world as the totality of all phenomena in space and time. Historically, it was often founded in religion; in modern use it addresses questions about the world and the universe which are beyond the scope of physical science. Cosmogony deals specifically with the origin of the universe, but the two concepts are closely related.

Pantheists, such as Spinoza, believe that God and the universe are one and the same. Panentheists, such as Plotinus, believe that the entire universe is part of God, but that God is greater than the universe. Deists, such as Voltaire, believe that God created the universe, set everything in motion, and then had nothing more to do with it. See the section on Philosophy of Religion for more details.

1.5 EPISTEMOLOGICAL FOUNDATION

Metaphysical study is conducted using deduction from that which is known a priori. Like foundational mathematics (which is sometimes considered a special case of metaphysics applied to the existence of number), it tries to give a coherent account of the structure of the world, capable of explaining our everyday and scientific perception of the world, and being free from contradictions. In mathematics, there are many different ways to define numbers; similarly, in metaphysics, there are many different ways to define objects, properties, concepts, and other entities that are claimed to make up the world. While metaphysics may, as a special case, study the entities postulated by fundamental science such as atoms and superstrings, its core topic is the set of categories such as object, property and causality which those scientific theories assume. For example: claiming that "electrons have charge" is a scientific theory; while exploring what it means for electrons to be (or at least, to be perceived as) "objects", charge to be a "property", and for both to exist in a topological entity called "space" is the task of metaphysics.

There are two broad stances about what is "the world" studied by metaphysics. According to metaphysical realism, the objects studied by metaphysics exist independently of any observer so that the subject is the most fundamental of all sciences. Metaphysical anti-realism, on the other hand, assumes that the objects studied by metaphysics exist inside the mind of an observer, so the subject becomes a form of introspection and conceptual analysis. This position is of more recent origin. Some philosophers, notably Kant, discuss both of these "worlds" and what can be inferred about each one. Some, such as the logical positivists, and many scientists, reject the metaphysical realism as meaningless and unverifiable. Others reply that this criticism also applies to any type of knowledge, including hard science, which claims to describe anything other than the contents of human perception, and thus that the world of perception is the objective world in some sense. Metaphysics itself usually assumes that some stance has been taken on these questions and that it may proceed independently of the choice—the question of which stance to take belongs instead to another branch of philosophy,

1.5.1 Central Questions

Ontology (being)

Ontology is the branch of philosophy that studies concepts such as existence, being, becoming, and reality. It includes the questions of how entities are grouped into basic categories and which of these entities exist on the most fundamental level. Ontology is sometimes referred to as the science of being. It has been characterized as general metaphysics in contrast to special metaphysics, which is concerned with more particular aspects of being. Ontologists often try to determine what the categories or highest kinds are and how they form a system of categories that provides an encompassing classification of all entities. Commonly proposed categories include substances, properties, relations, states of affairs and events. These categories are characterized by fundamental ontological concepts, like particularity and universality, abstractness and concreteness or possibility and necessity. Of special interest is the concept of ontological dependence, which determines whether the entities of a category exist on the most fundamental level. Disagreements within ontology are often about whether entities belonging to a certain category exist and, if so, how they are related to other entities.

Identity and Change

Identity is a fundamental metaphysical concern. Metaphysicians investigating identity are tasked with the question of what, exactly, it means for something to be identical to itself, or – more controversially – to something else. Issues of identity arise in the context of time: what does it mean for something to be itself across two moments in time? How do we account for this? Another question of identity arises when we ask what our criteria ought to be for determining identity, and how the reality of identity interfaces with linguistic expressions.

The metaphysical positions one takes on identity have far-reaching implications on issues such as the Mind–body problem, personal identity, ethics, and law.

A few ancient Greeks took extreme positions on the nature of change. Parmenides denied change altogether, while Heraclitus argued that change was ubiquitous: "No man ever steps in the same river twice."

Identity, sometimes called numerical identity, is the relation that a thing bears to itself, and which no thing bears to anything other than itself.

A modern philosopher who made a lasting impact on the philosophy of identity was Leibniz, whose Law of the Indiscernibility of Identicals is still widely accepted today. It states that if some object x is identical to some object y, then any property that x has, y will have as well.

Put formally, it states.

However, it does seem that objects can change over time. If one were to look at a tree one day, and the tree later lost a leaf, it would seem that one could still be looking at that same tree. Two rival theories to account for the relationship between change and identity are perdurantism, which treats the tree as a series of tree-stages, and endurantism, which maintains that the organism—the same tree—is present at every stage in its history.

By appealing to intrinsic and extrinsic properties, endurantism finds a way to harmonize identity with change. Endurantists believe that objects persist by being strictly numerically identical over time. However, if Leibniz's Law of the Indiscernibility of Identicals is utilized to define numerical identity here, it seems that objects must be completely unchanged in order to persist. Discriminating between intrinsic properties and extrinsic properties, endurantists state that numerical identity means that, if some object x is identical to some object y, then any intrinsic property that x has, y will have as well. Thus, if an object persists, intrinsic properties of it are unchanged, but extrinsic properties can change over time. Besides the object itself, environments and other objects can change over time; properties that relate to other objects would change even if this object does not change.

Perdurantism can harmonize identity with change in another way. In four-dimensionalism, a version of perdurantism, what persists is a four-dimensional object which does not change although three-dimensional slices of the object may differ.

Space and Time

Objects appear to us in space and time, while abstract entities such as classes, properties, and relations do not. How do space and time serve this function as a ground for objects? Are space and time entities themselves, of some form? Must they exist prior to objects? How exactly can they be defined? How is time related to change; must there always be something changing in order for time to exist?

Causality

Classical philosophy recognized a number of causes, including teleological future causes. In special relativity and quantum field theory the notions of space, time and causality become tangled together, with temporal orders of causations becoming dependent on who is observing them.[citation needed] The laws of physics are symmetrical in time, so could equally well be used to describe time as running backwards. Why then do we perceive it as flowing in one direction, the arrow of time, and as containing causation flowing in the same direction?

For that matter, can an effect precede its cause? This was the title of a 1954 paper by Michael Dummett, which sparked a discussion that continues today. Earlier, in 1947, C. S. Lewis had argued that one can meaningfully pray concerning the outcome of, e.g., a medical test while recognizing that the outcome is determined by past events: "My free act contributes to the cosmic shape." Likewise, some interpretations of quantum mechanics, dating to 1945, involve backward-in-time causal influences.

Causality is linked by many philosophers to the concept of counterfactuals. To say that A caused B means that if A had not happened then B would not have happened. This view was advanced by David Lewis in his 1973 paper "Causation". His subsequent papers further develop his theory of causation.

Causality is usually required as a foundation for philosophy of science if science aims to understand causes and effects and make predictions about them.

1.5.2 Necessity and Possibility

Metaphysicians investigate questions about the ways the world could have been. David Lewis, in On the Plurality of Worlds, endorsed a view called concrete modal realism, according to which facts about how things could have been are made true by other concrete worlds in which things are different. Other philosophers, including Gottfried Leibniz, have dealt with the idea of possible worlds as well. A necessary fact is true across all possible worlds. A possible fact is true in some possible world, even if not in the actual world. For example, it is possible that cats could have had two tails, or that any particular apple could have not existed. By contrast, certain propositions seem necessarily true, such as analytic propositions, e.g., "All bachelors are unmarried." The view that any analytic truth is necessary is not universally held among philosophers. A less controversial view is that self-identity is necessary, as it seems fundamentally incoherent to claim that any x is not identical to itself; this is known as the law of identity, a putative "first principle". Similarly, Aristotle describes the principle of non-contradiction:

> It is impossible that the same quality should both belong and not belong to the same thing... This is the most certain of all principles... Wherefore they who demonstrate refer to this as an ultimate opinion. For it is by nature the source of all the other axioms.

1.6 PERIPHERAL QUESTIONS

1.6.1 Metaphysical Cosmology and Cosmogony

Metaphysical cosmology is the branch of metaphysics that deals with the world as the totality of all phenomena in space and time. Historically, it formed a major part of the subject alongside Ontology, though its role is more peripheral in contemporary philosophy. It has had a broad scope, and in many cases was founded in religion. The ancient Greeks drew no distinction between this use and their model for the cosmos. However, in modern times it addresses questions about the Universe which are beyond the scope of the physical sciences. It is distinguished from religious cosmology in that it approaches these questions using philosophical methods.

Cosmogony deals specifically with the origin of the universe. Modern metaphysical cosmology and cosmogony try to address questions such as:

- What is the origin of the Universe? What is its first cause? Is its existence necessary?
- What are the ultimate material components of the Universe?
- What is the ultimate reason for the existence of the Universe? Does the cosmos have a purpose?

Mind and Matter

Accounting for the existence of mind in a world largely composed of matter is a metaphysical problem which is so large and important as to have become a specialized subject of study in its own right, philosophy of mind.

Substance dualism is a classical theory in which mind and body are essentially different, with the mind having some of the attributes traditionally assigned to the soul, and which creates an immediate conceptual puzzle about how the two interact. This form of substance dualism differs from the dualism of some eastern philosophical traditions (like Nyaya), which also posit a soul; for the soul, under their view, is ontologically distinct from the mind. Idealism postulates that material objects do not exist unless perceived and only as perceptions. Adherents of panpsychism, a kind of property dualism, hold that everything has a mental aspect, but not that everything exists in a mind. Neutral monism postulates that existence consists of a single substance that in itself is neither mental nor physical, but is capable of mental and physical aspects or attributes – thus it implies a dual-aspect theory. For the last century, the dominant theories have been science-inspired including materialistic monism, type identity theory, token identity theory, functionalism, reductive physicalism, nonreductive physicalism, eliminative materialism, anomalous monism, property dualism, epiphenomenalism and emergence.

Natural and Social Kinds

The earliest type of classification of social construction traces back to Plato in his dialogue Phaedrus where he claims that the biological classification system seems to carve nature at the joints. In contrast, later philosophers such as Michel Foucault and Jorge Luis Borges have challenged the capacity of natural and social classification. In his essay The Analytical Language of John Wilkins, Borges makes us imagine a certain encyclopedia where the animals are divided into (a) those that belong to the emperor; (b) embalmed ones; (c) those that are trained;... and so forth, in order to bring forward the ambiguity of natural and social kinds. According to metaphysics author Alyssa Ney: "the reason all this is interesting is that there seems to be a metaphysical difference between the Borgesian system and Plato's". The difference is not obvious but one classification attempts to carve entities up according to objective distinction while the other does not. According to Quine this notion is closely related to the notion of similarity.

Number

There are different ways to set up the notion of number in metaphysics theories. Platonist theories postulate number as a fundamental category itself. Others consider it to be a property of an entity called a "group" comprising other entities; or to be a relation held between several groups of entities, such as "the number four is the set of all sets of four things". Many of the debates around universals are applied to the study of number, and are of particular importance due to its status as a foundation for the philosophy of mathematics and for mathematics itself.

1.6.2 Applied Metaphysics

Although metaphysics as a philosophical enterprise is highly hypothetical, it also has practical application in most other branches of philosophy, science, and now also information technology. Such areas generally assume some basic ontology as well as other metaphysical stances on topics such as causality and agency, then build their own particular theories upon these.

In science, for example, some theories are based on the ontological assumption of objects with properties while others may reject objects completely (such as quantum field theories, where spread-out "electronness" becomes property of space-time rather than an object).

"Social" branches of philosophy such as philosophy of morality, aesthetics and philosophy of religion all require metaphysical foundations, which may be considered as branches or applications of metaphysics. For example, they may postulate the existence of basic entities such as value, beauty, and God. Then they use these postulates to make their own arguments about consequences resulting from them. When philosophers in these subjects make their foundations they are doing applied metaphysics, and may draw upon its core topics and methods to guide them, including ontology and other core and peripheral topics.

As in science, the foundations chosen will in turn depend on the underlying ontology used, so philosophers in these subjects may have to dig right down to the ontological layer of metaphysics to find what is possible for their theories. For example, a contradiction obtained in a theory of God or Beauty might be due to an assumption that it is an object rather than some other kind of ontological entity.

1.7 RELATION TO OTHER DISCIPLINES

1.7.1 Science

Prior to the modern history of science, scientific questions were addressed as a part of natural philosophy. Originally, the term "science" simply meant "knowledge". The scientific method, however, transformed natural philosophy into an empirical activity deriving from experiment, unlike the rest of philosophy. By the end of the 18th century, it had begun to be called "science" to distinguish it from other branches of philosophy. Science and philosophy have been considered separated disciplines ever since. Thereafter, metaphysics denoted philosophical enquiry of a non-empirical character into the nature of existence.

Metaphysics continues asking "why" where science leaves off. For example, any theory of fundamental physics is based on some set of axioms, which may postulate the existence of entities such as atoms, particles, forces, charges, mass, or fields. Stating such postulates is considered to be the "end" of a science theory. Metaphysics takes these postulates and explores what they mean as human concepts. For example, do all theories of physics require the existence of space and time, objects, and properties? Or can they be expressed using only objects, or only properties? Do the objects have to retain their identity over time or can they change? If they change, then are they still the same object? Can theories be reformulated by converting properties or predicates (such as "red") into entities or processes. Is the distinction between objects and properties fundamental to the physical world or to our perception of it?

Much recent work has been devoted to analyzing the role of metaphysics in scientific theorizing. Alexandre Koyré led this movement, declaring in his book Metaphysics and Measurement, "It is not by following experiment, but by outstripping experiment, that the scientific mind makes progress." That metaphysical propositions can influence scientific theorizing is John Watkins' most lasting contribution to philosophy. Since 1957 "he showed the ways in which some un-testable and hence, according to Popperian ideas, non-empirical propositions can nevertheless be influential in the development of properly testable and hence scientific theories. These profound results in applied elementary logic... represented an important corrective to positivist teachings about the meaninglessness of metaphysics and of normative claims". Imre Lakatos maintained that all scientific theories have a metaphysical "hard core" essential for the generation of hypotheses and theoretical assumptions. Thus, according to Lakatos, "scientific changes are connected with vast cataclysmic metaphysical revolutions."

An example from biology of Lakatos' thesis: David Hull has argued that changes in the ontological status of the species concept have been central in the development of biological thought from Aristotle through Cuvier, Lamarck, and Darwin. Darwin's ignorance of metaphysics made it more difficult for him to respond to his critics because he could not readily grasp the ways in which their underlying metaphysical views differed from his own.

In physics, new metaphysical ideas have arisen in connection with quantum mechanics, where subatomic particles arguably do not have the same sort of individuality as the particulars with which philosophy has traditionally been concerned. Also, adherence to a deterministic metaphysics in the face of the challenge posed by the quantum-mechanical uncertainty principle led physicists such as Albert Einstein to propose alternative theories that retained determinism. A.N. Whitehead is famous for creating a process philosophy metaphysics inspired by electromagnetism and special relativity.

In chemistry, Gilbert Newton Lewis addressed the nature of motion, arguing that an electron should not be said to move when it has none of the properties of motion.

Katherine Hawley notes that the metaphysics even of a widely accepted scientific theory may be challenged if it can be argued that the metaphysical presuppositions of the theory make no contribution to its predictive success.

Theology

There is a relationship between theological doctrines and philosophical reflection in the philosophy of a religion; philosophical reflections are strictly rational. On this way of seeing the two disciplines, if at least one of the premises of an argument is derived from revelation, the argument falls in the domain of theology; otherwise it falls into philosophy's domain.

1.7.2 Rejections of Metaphysics

Meta-metaphysics is the branch of philosophy that is concerned with the foundations of metaphysics. A number of individuals have suggested that much or all of metaphysics should be rejected, a meta-metaphysical position known as metaphysical deflationism[a] or ontological deflationism.

In the 16th century, Francis Bacon rejected scholastic metaphysics, and argued strongly for what is now called empiricism, being seen later as the father of modern empirical science. In the 18th century, David Hume took a strong position, arguing that all genuine knowledge involves either mathematics or matters of fact and that metaphysics, which goes beyond these, is worthless. He concludes his Enquiry Concerning Human Understanding (1748) with the statement:

If we take in our hand any volume; of divinity or school metaphysics, for instance; let us ask, Does it contain any abstract reasoning concerning quantity or number? No. Does it contain any experimental reasoning concerning matter of fact and existence? No. Commit it then to the flames: for it can contain nothing but sophistry and illusion.

Thirty-three years after Hume's Enquiry appeared, Immanuel Kant published his Critique of Pure Reason. Although he followed Hume in rejecting much of previous metaphysics, he argued that there was still room for some synthetic a priori knowledge, concerned with matters of fact yet obtainable independent of experience. These included fundamental structures of space, time, and causality. He also argued for the freedom of the will and the existence of "things in themselves", the ultimate objects of experience.

Wittgenstein introduced the concept that metaphysics could be influenced by theories of aesthetics, via logic, vis. a world composed of "atomical facts".

In the 1930s, A.J. Ayer and Rudolf Carnap endorsed Hume's position; Carnap quoted the passage above. They argued that metaphysical statements are neither true nor false but meaningless since, according to their verifiability theory of meaning, a statement is meaningful only if there can be empirical evidence for or against it. Thus, while Ayer rejected the monism of Spinoza, he avoided a commitment to pluralism, the contrary position, by holding both views to be without meaning. Carnap took a similar line with the controversy over the reality of the external world. While the logical positivism movement is now considered dead, it has continued to influence philosophy development.

Arguing against such rejections, the Scholastic philosopher Edward Feser held that Hume's critique of metaphysics, and specifically Hume's fork, is "notoriously self-refuting". Feser argues that Hume's fork itself is not a conceptual truth and is not empirically testable.

Some living philosophers, such as Amie Thomasson, have argued that many metaphysical questions can be dissolved just by looking at the way we use words; others, such as Ted Sider, have argued that metaphysical questions are substantive, and that we can make progress toward answering them by comparing theories according to a range of theoretical virtues inspired by the sciences, such as simplicity and explanatory power.

Etymology

The word "metaphysics" derives from the Greek words μετά (metá, "after") and φυσικά (physiká, "physics"). It was first used as the title for several of Aristotle's works, because they were usually anthologized after the works on physics in complete editions. The prefix meta- ("after") indicates that these works come "after" the chapters on physics. However, Aristotle himself did not call the subject of these books metaphysics: he referred to it as "first philosophy".

However, once the name was given, the commentators sought to find other reasons for its appropriateness. For instance, Thomas Aquinas understood it to refer to the chronological or pedagogical order among our philosophical studies, so that the "metaphysical sciences" would mean "those that we study after having mastered the sciences that deal with the physical world".

The term was misread by other medieval commentators, who thought it meant "the science of what is beyond the physical". Following this tradition, the prefix meta- has more recently been prefixed to the names of sciences to designate higher sciences dealing with ulterior and more fundamental problems: hence metamathematics, metaphysiology, etc.

A person who creates or develops metaphysical theories is called a metaphysician.

Common parlance also uses the word "metaphysics" for a different referent from that of the present article, namely for beliefs in arbitrary non-physical or magical entities. For example, "Metaphysical healing" to refer to healing by means of remedies that are magical rather than scientific. This usage stemmed from the various historical schools of speculative metaphysics which operated by postulating all manner of physical, mental and spiritual entities as bases for particular metaphysical systems. Metaphysics as a subject does not preclude beliefs in such magical entities but neither does it promote them. Rather, it is the subject which provides the vocabulary and logic with which such beliefs might be analyzed and studied, for example to search for inconsistencies both within themselves and with other accepted systems such as Science.

1.8 HISTORY AND SCHOOLS OF METAPHYSICS

1.8.1 Pre-History

Cognitive archeology such as analysis of cave paintings and other pre-historic art and customs suggests that a form of perennial philosophy or Shamanic metaphysics may stretch back to the birth of behavioral modernity, all around the world. Similar beliefs are found in present-day "stone age" cultures such as Australian aboriginals. Perennial philosophy postulates the existence of a spirit or concept world alongside the day-to-day world, and interactions between these worlds during dreaming and ritual, or on special days or at special places. It has been argued that perennial philosophy formed the basis for Platonism, with Plato articulating, rather than creating, much older widespread beliefs.

1.8.2 Bronze Age

Bronze Age cultures such as ancient Mesopotamia and ancient Egypt (along with similarly structured but chronologically later cultures such as Mayans and Aztecs) developed belief systems based on mythology, anthropomorphic gods, mind–body dualism, and a spirit world,[citation needed] to explain causes and cosmology. These cultures appear to have been interested in astronomy and may have associated or identified the stars with some of these entities. In ancient Egypt, the ontological distinction between order (maat) and chaos (Isfet) seems to have been important.

1.8.3 Pre-Socratic Greece

The first named Greek philosopher, according to Aristotle, is Thales of Miletus, early 6th century BCE. He made use of purely physical explanations to explain the phenomena of the world rather than the mythological and divine explanations of tradition. He is thought to have posited water as the single underlying principle (or arche in later Aristotelian terminology) of the material world. His fellow, but younger Miletians, Anaximander and Anaximenes, also posited monistic underlying principles, namely apeiron (the indefinite or boundless) and air respectively.

Another school was the Eleatics, in southern Italy. The group was founded in the early fifth century BCE by Parmenides, and included Zeno of Elea and Melissus of Samos. Methodologically, the Eleatics were broadly rationalist, and took logical standards of clarity and necessity to be the criteria of truth. Parmenides' chief doctrine was that reality is a single unchanging and universal Being. Zeno used reductio ad absurdum, to demonstrate the illusory nature of change and time in his paradoxes.

Heraclitus of Ephesus, in contrast, made change central, teaching that "all things flow". His philosophy, expressed in brief aphorisms, is quite cryptic. For instance, he also taught the unity of opposites.

Democritus and his teacher Leucippus, are known for formulating an atomic theory for the cosmos. They are considered forerunners of the scientific method.

1.8.4 Classical China

Metaphysics in Chinese philosophy can be traced back to the earliest Chinese philosophical concepts from the Zhou Dynasty such as Tian (Heaven) and Yin and Yang. The fourth century BCE saw a turn towards cosmogony with the rise of

Taoism (in the Daodejing and Zhuangzi) and sees the natural world as dynamic and constantly changing processes which spontaneously arise from a single immanent metaphysical source or principle (Tao).

Another philosophical school which arose around this time was the School of Naturalists which saw the ultimate metaphysical principle as the Taiji, the "supreme polarity" composed of the forces of Yin and Yang which were always in a state of change seeking balance. Another concern of Chinese metaphysics, especially Taoism, is the relationship and nature of Being and non-Being. The Taoists held that the ultimate, the Tao, was also non-being or no-presence. Other important concepts were those of spontaneous generation or natural vitality (Ziran) and "correlative resonance" (Ganying).

After the fall of the Han Dynasty (220 CE), China saw the rise of the Neo-Taoist Xuanxue school. This school was very influential in developing the concepts of later Chinese metaphysics. Buddhist philosophy entered China (c. 1st century) and was influenced by the native Chinese metaphysical concepts to develop new theories. The native Tiantai and Huayen schools of philosophy maintained and reinterpreted the Indian theories of shunyata and Buddha-nature into the theory of interpenetration of phenomena. Neo-Confucians like Zhang Zai under the influence of other schools developed the concepts of "principle" (li) and vital energy (qi).

1.9 CLASSICAL GREECE

1.9.1 Socrates and Plato

Socrates is known for his dialectic or questioning approach to philosophy rather than a positive metaphysical doctrine.

His pupil, Plato is famous for his theory of forms. Platonic realism is considered to be a solution to the problem of universals; i.e., what particular objects have in common is that they share a specific Form which is universal to all others of their respective kind.

The theory has a number of other aspects:

- Epistemological: knowledge of the Forms is more certain than mere sensory data.
- Ethical: The Form of the Good sets an objective standard for morality.
- Time and Change: The world of the Forms is eternal and unchanging. Time and change belong only to the lower sensory world. "Time is a moving image of Eternity".
- Abstract objects and mathematics: Numbers, geometrical figures, etc., exist mind-independently in the World of Forms.

Platonism developed into Neoplatonism, a philosophy with a monotheistic and mystical flavour that survived well into the early Christian era.

1.9.2 Aristotle

Plato's pupil Aristotle wrote widely on almost every subject, including metaphysics. His solution to the problem of universals contrasts with Plato's. Whereas Platonic Forms are existentially apparent in the visible world, Aristotelian essences dwell in particulars.

Potentiality and Actuality are principles of a dichotomy which Aristotle used throughout his philosophical works to analyze motion, causality and other issues.

The Aristotelian theory of change and causality stretches to four causes: the material, formal, efficient and final. The efficient cause corresponds to what is now known as a cause simplicity. Final causes are explicitly teleological, a concept now regarded as controversial in science. The Matter/Form dichotomy was to become highly influential in later philosophy as the substance/essence distinction.

The opening arguments in Aristotle's Metaphysics, Book I, revolve around the senses, knowledge, experience, theory, and wisdom. The first main focus in the Metaphysics is attempting to determine how intellect "advances from sensation through memory, experience, and art, to theoretical knowledge". Aristotle claims that eyesight provides us with the capability to recognize and remember experiences, while sound allows us to learn.

1.9.3 Classical India

Samkhya

Samkhya is an ancient system of Indian philosophy based on a dualism involving the ultimate principles of consciousness and matter. It is described as the rationalist school of Indian philosophy. It is most related to the Yoga school of Hinduism, and its method was most influential on the development of Early Buddhism.

The Samkhya is an enumerationist philosophy whose epistemology accepts three of six pramanas (proofs) as the only reliable means of gaining knowledge. These include pratyaksa (perception), anumana (inference) and śabda (aptavacana, word/testimony of reliable sources).

Samkhya is strongly dualist. Samkhya philosophy regards the universe as consisting of two realities; purusa (consciousness) and prakrti (matter). Jiva (a living being) is that state in which purusa is bonded to prakrti in some form. This fusion, state the Samkhya scholars, led to the emergence of buddhi ("spiritual awareness") and ahankara (ego consciousness). The universe is described by this school as one created by purusa-prakrti entities infused with various permutations and combinations of variously enumerated elements, senses, feelings, activity and mind. During the state of imbalance, one of more constituents overwhelm the others, creating a form of bondage, particularly of the mind. The end of this imbalance, bondage is called liberation, or moksha, by the Samkhya school.

The existence of God or supreme being is not directly asserted, nor considered relevant by the Samkhya philosophers. Samkhya denies the final cause of Ishvara (God). While the Samkhya school considers the Vedas as a reliable source of knowledge, it is an atheistic philosophy according to Paul Deussen and other scholars. A key difference between Samkhya and Yoga schools, state scholars, is that Yoga school accepts a "personal, yet essentially inactive, deity" or "personal god".

Samkhya is known for its theory of gunas (qualities, innate tendencies). Guna, it states, are of three types: sattva being good, compassionate, illuminating, positive, and constructive; rajas is one of activity, chaotic, passion, impulsive, potentially good or bad; and tamas being the quality of darkness, ignorance, destructive, lethargic, negative. Everything, all life forms and human beings, state Samkhya scholars, have these three gunas, but in different proportions.

The interplay of these gunas defines the character of someone or something, of nature and determines the progress of life. The Samkhya theory of gunas was widely discussed, developed and refined by various schools of Indian philosophies, including Buddhism. Samkhya's philosophical treatises also influenced the development of various theories of Hindu ethics.

Vedanta

Realization of the nature of Self-identity is the principal object of the Vedanta system of Indian metaphysics. In the Upanishads, self-consciousness is not the first-person indexical self-awareness or the self-awareness which is self-reference without identification, and also not the self-consciousness which as a kind of desire is satisfied by another self-consciousness. It is Self-realisation; the realisation of the Self consisting of consciousness that leads all else.

The word self-consciousness in the Upanishads means the knowledge about the existence and nature of manusya, human being. It means the consciousness of our own real being, the primary reality. Self-consciousness means self-knowledge, the knowledge of Prajna i.e. of Prana which is attained by a Brahman. According to the Upanishads the Atman or Paramatman is phenomenally unknowable; it is the object of realisation. The Atman is unknowable in its essential nature; it is unknowable in its essential nature because it is the eternal subject who knows about everything including itself. The Atman is the knower and also the known.

Metaphysicians regard the Self either to be distinct from the Absolute or entirely identical with the Absolute. They have given form to three schools of thought – (a) the Dualistic school, (b) the Quasi-dualistic school and c) the Monistic school, as the result of their varying mystical experiences. Prakrti and Atman, when treated as two separate and distinct aspects form the basis of the Dualism of the Shvetashvatara Upanishad. Quasi-dualism is reflected in the Vaishnavite-monotheism of Ramanuja and the absolute Monism, in the teachings of Adi Shankara.

Self-consciousness is the Fourth state of consciousness or Turiya, the first three being Vaisvanara, Taijasa and Prajna. These are the four states of individual consciousness.

There are three distinct stages leading to Self-realisation. The First stage is in mystically apprehending the glory of the Self within us as though we were distinct from it. The Second stage is in identifying the "I-within" with the Self, that we are in essential nature entirely identical with the pure Self. The Third stage is in realising that the Atman is Brahman, that there is no difference between the Self and the Absolute. The Fourth stage is in realising "I am the Absolute" –Aham Brahman Asmi. The Fifth stage is in realising that Brahman is the "All" that exists, as also that which does not exist.

1.10 BUDDHIST METAPHYSICS

In Buddhist philosophy there are various metaphysical traditions that have proposed different questions about the nature of reality based on the teachings of the Buddha in the early Buddhist texts. The Buddha of the early texts does not focus on metaphysical questions but on ethical and spiritual training and in some cases, he dismisses certain metaphysical questions as unhelpful and indeterminate Avyakta, which he recommends should be set aside. The development of systematic metaphysics arose after the Buddha's death with the rise of the Abhidharma traditions. The Buddhist Abhidharma schools developed their analysis of reality based on the concept of dharmas which are the ultimate physical and mental events that makeup experience and their relations to each other. Noa Ronkin has called their approach "phenomenological".

Later philosophical traditions include the Madhyamika school of Nagarjuna, which further developed the theory of the emptiness (shunyata) of all phenomena or dharmas which rejects any kind of substance. This has been interpreted as a form of anti-foundationalism and anti-realism which sees reality as having no ultimate essence or ground. The Yogacara school meanwhile promoted a theory called "awareness only" (vijnapti-matra) which has been interpreted as a form of Idealism or Phenomenology and denies the split between awareness itself and the objects of awareness.

1.11 ISLAMIC METAPHYSICS

Major ideas in Sufi metaphysics have surrounded the concept of wehdah meaning "unity". The phrase has been translated "pantheism." Wujud (i.e. existence or presence) here refers to Allah's wujud (compare tawhid). On the other hand, wahdat ash-shuhūd, meaning "Apparentism" or "Monotheism of Witness", holds that God and his creation are entirely separate.

1.11.1 Scholasticism and the Middle Ages

Between about 1100 and 1500, philosophy as a discipline took place as part of the Catholic church's teaching system, known as scholasticism. Scholastic philosophy took place within an established framework blending Christian theology with Aristotelian teachings. Although fundamental orthodoxies were not commonly challenged, there were nonetheless deep metaphysical disagreements, particularly over the problem of universals, which engaged Duns Scotus and Pierre Abelard. William of Ockham is remembered for his principle of ontological parsimony.

Continental Rationalism

In the early modern period (17th and 18th centuries), the system-building scope of philosophy is often linked to the rationalist method of philosophy, that is the technique of deducing the nature of the world by pure reason. The scholastic concepts of substance and accident were employed.

- Leibniz proposed in his Monadology a plurality of non-interacting substances.
- Descartes is famous for his dualism of material and mental substances.
- Spinoza believed reality was a single substance of God-or-nature.

Wolff

Christian Wolff had theoretical philosophy divided into an ontology or philosophia prima as a general metaphysics, which arises as a preliminary to the distinction of the three "special metaphysics" on the soul, world and God: rational psychology, rational cosmology and rational theology. The three disciplines are called empirical and rational because they are independent of revelation. This scheme, which is the counterpart of religious tripartition in creature, creation, and Creator, is best known to philosophical students by Kant's treatment of it in the Critique of Pure Reason. In the "Preface" of the 2nd edition of Kant's book, Wolff is defined "the greatest of all dogmatic philosophers."

British Empiricism

British empiricism marked something of a reaction to rationalist and system-building metaphysics, or speculative metaphysics as it was pejoratively termed. The skeptic David Hume famously declared that most metaphysics should be consigned to the flames. Hume was notorious among his contemporaries as one of the first philosophers to openly doubt religion, but is better known now for his critique of causality. John Stuart Mill, Thomas Reid and John Locke were less skeptical, embracing a more cautious style of metaphysics based on realism, common sense and science. Other philosophers, notably George Berkeley were led from empiricism to idealistic metaphysics.

Kant

Immanuel Kant attempted a grand synthesis and revision of the trends already mentioned: scholastic philosophy, systematic metaphysics, and skeptical empiricism, not to forget the burgeoning science of his day. As did the systems builders, he had an overarching framework in which all questions were to be addressed. Like Hume, who famously woke him from his 'dogmatic slumbers', he was suspicious of metaphysical speculation, and also places much emphasis on the limitations of the human mind. Kant described his shift in metaphysics away from making claims about an objective noumenal world, towards exploring the subjective phenomenal world, as a Copernican Revolution, by analogy to (though opposite in direction to) Copernicus' shift from man (the subject) to the sun (an object) at the center of the universe.

Kant saw rationalist philosophers as aiming for a kind of metaphysical knowledge he defined as the synthetic apriori—that is knowledge that does not come from the senses (it is a priori) but is nonetheless about reality (synthetic). Inasmuch as it is about reality, it differs from abstract mathematical propositions (which he terms analytical apriori), and being apriori it is distinct from empirical, scientific knowledge (which he terms synthetic aposteriori). The only synthetic apriori knowledge we can have is of how our minds organise the data of the senses; that organising framework is space and time, which for Kant have no mind-independent existence, but nonetheless operate uniformly in all humans. Apriori knowledge of space and time is all that remains of metaphysics as traditionally conceived. There is a reality beyond sensory data or phenomena, which he calls the realm of noumena; however, we cannot know it as it is in itself, but only as it appears to us. He allows himself to speculate that the origins of phenomenal God, morality, and free will might exist in the noumenal realm, but these possibilities have to be set against its basic unknowability for humans. Although he saw himself as having disposed of metaphysics, in a sense, he has generally been regarded in retrospect as having a metaphysics of his own, and as beginning the modern analytical conception of the subject.

1.11.2 Late Modern Philosophy

Nineteenth century philosophy was overwhelmingly influenced by Kant and his successors. Schopenhauer, Schelling, Fichte and Hegel all purveyed their own panoramic versions of German Idealism, Kant's own caution about metaphysical speculation, and refutation of idealism, having fallen by the wayside. The idealistic impulse continued into the early twentieth century with British idealists such as F. H. Bradley and J. M. E. McTaggart. Followers of Karl Marx took Hegel's dialectic view of history and re-fashioned it as materialism.

1.11.3 Early Analytic Philosophy and Positivism

During the period when idealism was dominant in philosophy, science had been making great advances. The arrival of a new generation of scientifically minded philosophers led to a sharp decline in the popularity of idealism during the 1920s.

Analytic philosophy was spearheaded by Bertrand Russell and G. E. Moore. Russell and William James tried to compromise between idealism and materialism with the theory of neutral monism.

The early to mid-twentieth-century philosophy saw a trend to reject metaphysical questions as meaningless. The driving force behind this tendency was the philosophy of logical positivism as espoused by the Vienna Circle, which argued that the meaning of a statement was its prediction of observable results of an experiment, and thus that there is no need to postulate the existence of any objects other than these perceptual observations.

At around the same time, the American pragmatists were steering a middle course between materialism and idealism. System-building metaphysics, with a fresh inspiration from science, was revived by A. N. Whitehead and Charles Hartshorne.

Continental Philosophy

The forces that shaped analytic philosophy—the break with idealism, and the influence of science—were much less significant outside the English speaking world, although there was a shared turn toward language. Continental philosophy continued in a trajectory from post Kantianism.

The phenomenology of Husserl and others was intended as a collaborative project for the investigation of the features and structure of consciousness common to all humans, in line with Kant's basing his synthetic apriori on the uniform operation of consciousness. It was officially neutral with regards to ontology, but was nonetheless to spawn a number of metaphysical systems. Brentano's concept of intentionality would become widely influential, including on analytic philosophy.

Heidegger, author of Being and Time, saw himself as re-focusing on Being-qua-being, introducing the novel concept of Dasein in the process. Classing himself an existentialist, Sartre wrote an extensive study of Being and Nothingness.

The speculative realism movement marks a return to full blooded realism.

Process Metaphysics

There are two fundamental aspects of everyday experience: change and persistence. Until recently, the Western philosophical tradition has arguably championed substance and persistence, with some notable exceptions, however. According to process thinkers, novelty, flux and accident do matter, and sometimes they constitute the ultimate reality.

In a broad sense, process metaphysics is as old as Western philosophy, with figures such as Heraclitus, Plotinus, Duns Scotus, Leibniz, David Hume, Georg Wilhelm Friedrich Hegel, Friedrich Wilhelm Joseph von Schelling, Gustav Theodor Fechner, Friedrich Adolf Trendelenburg, Charles Renouvier, Karl Marx, Ernst Mach, Friedrich Wilhelm Nietzsche, Émile Boutroux, Henri Bergson, Samuel Alexander and Nicolas Berdyaev. It seemingly remains an open question whether major "Continental" figures such as the late Martin Heidegger, Maurice Merleau-Ponty, Gilles Deleuze, Michel Foucault, or Jacques Derrida should be included.

In a strict sense, process metaphysics may be limited to the works of a few philosophers: G. W. F. Hegel, Charles Sanders Peirce, William James, Henri Bergson, A. N. Whitehead, and John Dewey. From a European perspective, there was a very significant and early Whiteheadian influence on the works of outstanding scholars such as Émile Meyerson (1859–1933), Louis Couturat (1868–1914), Jean Wahl (1888–1974), Robin George Collingwood (1889–1943), Philippe Devaux (1902–1979), Hans Jonas

(1903–1993), Dorothy M. Emmett (1904–2000), Maurice Merleau Ponty (1908–1961), Enzo Paci (1911–1976), Charlie Dunbar Broad (1887–1971), Wolfe Mays (1912–2005), Ilya Prigogine (1917–2003), Jules Vuillemin (1920–2001), Jean Ladrière (1921–2007), Gilles Deleuze (1925–1995), Wolfhart Pannenberg (1928–2014), and Reiner Wiehl (1929–2010).

Contemporary Analytic Philosophy

While early analytic philosophy tended to reject metaphysical theorizing, under the influence of logical positivism, it was revived in the second half of the twentieth century. Philosophers such as David K. Lewis and David Armstrong developed elaborate theories on a range of topics such as universals, causation, possibility and necessity and abstract objects. However, the focus of analytic philosophy generally is away from the construction of all-encompassing systems and toward close analysis of individual ideas.

Among the developments that led to the revival of metaphysical theorizing were Quine's attack on the analytic–synthetic distinction, which was generally taken to undermine Carnap's distinction between existence questions internal to a framework and those external to it.

The philosophy of fiction, the problem of empty names, and the debate over existence's status as a property have all come of relative obscurity into the limelight, while perennial issues such as free will, possible worlds, and the philosophy of time have had new life breathed into them.

The analytic view is of metaphysics as studying phenomenal human concepts rather than making claims about the noumenal world, so its style often blurs into philosophy of language and introspective psychology. Compared to system-building, it can seem very dry, stylistically similar to computer programming, mathematics or even accountancy.

2

MAJOR DOCTRINES

2.1 INTRODUCTION

Metaphysics is the branch of philosophy concerned with the nature of existence, being and the world. Arguably, metaphysics is the foundation of philosophy: Aristotle calls it "first philosophy" (or sometimes just "wisdom"), and says it is the subject that deals with "first causes and the principles of things".

It asks questions like: "What is the nature of reality?", "How does the world exist, and what is its origin or source of creation?", "Does the world exist outside the mind?", "How can the incorporeal mind affect the physical body?", "If things exist, what is their objective nature?", "Is there a God (or many gods, or no god at all)?"

Originally, the Greek word "metaphysika" (literally "after physics") merely indicated that part of Aristotle's oeuvre which came, in its sequence, after those chapters which dealt with physics. Later, it was misinterpreted by Medieval commentators on the classical texts as that which is above or beyond the physical, and so over time metaphysics has effectively become the study of that which transcends physics.

Aristotle originally split his metaphysics into three main sections and these remain the main branches of metaphysics:

- Ontology (the study of being and existence, including the definition and classification of entities, physical or mental, the nature of their properties, and the nature of change)
- Natural Theology (the study of God, including the nature of religion and the world, existence of the divine, questions about the creation, and the various other religious or spiritual issues)
- Universal Science (the study of first principles of logic and reasoning, such as the law of noncontradiction)

Metaphysics has been attacked, at different times in history, as being futile and overly vague, particularly by David Hume, Immanuel Kant and A.J. Ayer. It may be more useful to say that a metaphysical statement usually implies an idea about the world or the universe, which may seem reasonable but is ultimately not empirically verifiable, testable or provable.

Derived from the Greek meta ta physika ("after the things of nature"); referring to an idea, doctrine, or posited reality outside of human sense perception. In modern philosophical terminology, metaphysics refers to the studies of what cannot be reached through objective studies of material reality. Areas of metaphysical studies include ontology, cosmology, and often, epistemology.

Metaphysical - Longer definition: Metaphysics is a type of philosophy or study that uses broad concepts to help define reality and our understanding of it. Metaphysical studies generally seek to explain inherent or universal elements of reality which are not easily discovered or experienced in our everyday life. As such, it is concerned with explaining the features of reality that exist beyond the physical world and our immediate senses. Metaphysics, therefore, uses logic based on the meaning of human terms, rather than on a logic tied to human sense perception of the objective world. Metaphysics might include the study of the nature of the human mind, the definition and meaning of existence, or the nature of space, time, and/or causality.

The origin of philosophy, beginning with the Pre-Socratics, was metaphysical in nature. For example, the philosopher Plotinus held that the reason in the world and in the rational human mind is only a reflection of a more universal and perfect reality beyond our limited human reason. He termed this ordering power in the universe "God."

Metaphysical ideas, because they are not based on direct experience with material reality, are often in conflict with the modern sciences. Beginning with the Enlightenment and the Scientific Revolution, experiments with, and observations of, the world became the yardsticks for measuring truth and reality. Therefore, our contemporary valuation of scientific knowledge over other forms of knowledge helps explain the controversy and skepticism concerning metaphysical claims, which are considered unverifiable by modern science.

In matters of religion, the problem of validating metaphysical claims is most readily seen in all of the "proofs" for the existence of God. Like trying to prove the existence of a "soul" or "spirit" in the human, attempts to scientifically prove the existence of God and other nonobjective, nonhuman realities is seemingly impossible. The difficulty arises out of the attempt to scientifically study and objectify something which, by its very nature, cannot become an object of our scientific studies. This reigning belief that everything can be explained scientifically in terms of natural causes - referred to as naturalism - compels many to think that only what is seen or sensed, only what can be hypothesized and tested can be true, and therefore, meaningful to us as humans.

Recently, however, even as metaphysics has come under attack for its apparent lack of access to real knowledge, so has science begun to have its own difficulties in claiming absolute knowledge. Continual developments in our understanding of the human thought process reveals that science cannot solely be relied upon to explain reality, for the human mind cannot be seen as simply a mirror of the natural world. For example, since the act of scientific observation itself tends to produce the reality it hopes to explain, the so-called "truths" of science cannot be considered as final or objective. This fact manifests itself over and over again, as scientific truths and laws continue to break down or yield to new and better explanations of reality. What becomes apparent, therefore, is that the process of human interpretation in the sciences, as elsewhere, is both variable and relative to the observer's viewpoint.

Under the skeptical analyses of the philosophical movements known as postmodernism and deconstructionism, all of these facts have resulted in a modern repudiation of both metaphysics and science. Their criticisms are based on the cultural and historical relativity of all knowledge. These two philosophical "schools" deny any existence at all of an objective or universal knowledge. Thus, metaphysical claims stand today between the absolutist claims of science (scientism) and the complete relativism of postmodernism and deconstructionism.

Under the heading of Metaphysics, the major doctrines or theories include:

Agnosticism

Atheism

Atomism

Deism

Determinism

Dualism

Essentialism

Existentialism

Fideism

Idealism

Intellectualism

Materialism

Monism

Monotheism

Naturalism

Nominalism

Nihilism
Objectivism
Panentheism
Pantheism
Phenomenology
Physicalism
Pluralism
Polytheism
Realism
Reductionism
Relativism
Solipsism
Subjectivism
Theism
Voluntarism

2.2 PROBLEMS IN METAPHYSICS

2.2.1 The Reality of the External World

The Problem in Early Modern Philosophy

Although sensations (i.e., the conscious experiences that result from stimulation of the sense organs) are mental events, they seem to most people to be a source of information—fallible, perhaps, but in the main reliable—about a nonmental world, the world of material or physical objects, which constitutes the environment of the perceiver. Regarding that "external world," many philosophers have attempted to answer the following related questions: Is there an external world? If there is, do the senses provide reliable information about it? If they do, do human beings know—or can they come to know—what the external world is like? If they can, what exactly is the source or basis of that knowledge? To attempt to answer such questions is to address the problem of the reality of the external world.

That problem belongs entirely to modern (that is, postmedieval) philosophy; no ancient or medieval philosopher so much as considered any of the questions mentioned in the previous paragraph. First explored by the French philosopher René Descartes (1596–1650), it was not regarded as fundamental or especially important—i.e., as a problem that every philosophical system with any pretense to comprehensiveness was obliged to address—until the work of the Anglo-Irish philosopher George Berkeley (1685–1753) became widely known. Berkeley devised very able and ingenious arguments for a thoroughgoing form of idealism, according to which nothing exists but ideas (i.e., sensations and mental images), things composed of ideas, and the minds within which ideas exist. Although few philosophers accepted Berkeley's doctrine—his arguments were notorious rather than famous—it was generally considered important that it should be refuted. The typical attitude of philosophers of the 18th century to the problem of the reality of the external world was well summarized by the German Enlightenment philosopher Immanuel Kant (1724–1804), who wrote:

> It remains a scandal to philosophy, and to human reason in general, that it is necessary to take the existence of things outside us…merely on faith, and, if anyone should happen to doubt it, no adequate proof can be produced to oppose him.

Descartes's formulation of the problem is presented in his Meditationes de Prima Philosophia—a record, in the form of a first-person narrative, of its author's search for an absolutely reliable foundation of human knowledge. Descartes assumed that each person, in order to have knowledge, required such a foundation, and he further assumed that the foundation of each person's knowledge would be a set of propositions whose truth it was impossible for that person to doubt and on the basis of which additional propositions could be inferred. He proposed to demonstrate to each reader of the Meditations how that reader could find such a set of propositions by using himself, Descartes, as an example. He therefore set out to identify those propositions that it was impossible for him to doubt. Those propositions, Descartes argued, were precisely those that he could be certain of even in the following worst-case scenario: "I will suppose that…some malicious demon of the utmost power and cunning has employed all his energies to deceive me." In such a case, Descartes decided, what he could be certain of would be only his own present existence as a thinking, sensing being and his present thoughts and sensations. Such was the foundation upon which the edifice of his knowledge was to be constructed. The "ground floor" of the building would be a proof that the sensations that he had found within himself—sensations that seemed to him to represent physical or material objects like the body he animated and the pen in his fingers and the paper and writing table before him—were veridical. If Descartes's sensations were veridical, then an external world would exist, since "an external world" is no more than a name for the totality of objects of the kind whose existence Descartes's sensations testified to. And if Descartes knew that his sensations were veridical, he would know that there was an external world, and he would know that it was more or less as his sensations indicated that it was.

The central argument of the Meditations aims to establish the conclusion that its author's sensations are veridical. The core of the argument consists of two subordinate arguments, each of which seeks to prove the existence of a perfect being—that is, of God. The purpose of the current discussion does not require an exposition of those arguments; it suffices to point out that Descartes maintained that each of the subordinate arguments is valid and that one can recognize their validity without having to assume that one's sensations are veridical.

Having demonstrated (to his own satisfaction) the existence of a perfect being, Descartes proceeded to argue that if his (Descartes's) sensations were not veridical, the perfect being just established—the creator of all things and hence of all sensations—would be a deceiver. His argument concludes with the observation that deceit is an imperfection and hence incompatible with the nature of a perfect being. The hypothesis that his sensations are not veridical thus leads to a contradiction, from which it follows that his sensations are veridical after all. Descartes assumes that each of his readers will be able to establish the same conclusion with respect to the reader's own sensations.

Very few philosophers have found the argument of the Meditations convincing, if for no other reason than that very few of them have been convinced by either of the subordinate arguments for the existence of God. Moreover, very few philosophers have supposed that some other argument for the veridicality of sensations or for the existence of an external world succeeds where Descartes's argument fails.

Berkeleyan Idealism

Although Berkeley denied the existence of anything besides ideas, things composed of ideas, and the minds within which ideas exist, he did not explicitly deny the existence of objects such as "the body Descartes animated," "the pen in his fingers," and "the paper and writing table before him." Berkeley instead affirmed the existence of those objects but insisted that they were composed of ideas. His argument was straightforward. Adapted to the present case, its central point can be formulated as follows:

Take the whiteness of the sheet of paper before Descartes—or its (perceived) shape. Those properties cannot exist outside the mind—that is, outside anyone's mind. Thus, the properties of a sheet of paper exist in the mind. Furthermore, there is nothing more to a thing than its properties—in particular, there is no imperceptible "substrate" in which properties "inhere," as the English philosopher John Locke (1632–1704) held. (If there were such a thing, how would anyone know about it, given that all knowledge is derived from perception, and the substrate is by definition imperceptible?) Therefore, sheets of paper and all other objects of the kind that philosophers call "material" exist only in the mind. Indeed, there is nothing wrong with anyone's calling the sheet of paper material, if by that word the speaker means that it is extended in space and impenetrable (it cannot be penetrated by another extended object without suffering damage). But extension and impenetrability are properties like any other, and as such they exist only in the mind.

In a sense, therefore, Berkeley accepted the existence of an external world, because he affirmed the existence of "material" objects, of which the external world is ordinarily said to be composed. Moreover, for Berkeley, there are things that exist partly or wholly outside each person's mind—namely, ideas that are not in that person's mind, and minds that are not identical to that person's mind. But most philosophers who speak of the problem of the reality of the external world would deny that Berkeley's idealism is consistent with the reality of an external world as they understand that phrase. Kant is a case in point. His proof of the existence of "things outside us" occurs, in the Critique of Pure Reason, under the heading "Refutation of Idealism," and by "idealism" he meant Berkeleyan idealism (or "dogmatic" idealism, as he also called it—that is, an idealism that is defended on the basis of metaphysical reasoning of the kind that Kant's Critique was designed to refute). Indeed, almost all philosophers who have used the phrase "external world" would join Kant in regarding Berkeleyan idealism as inconsistent with the reality of an external world. Specifically, they would insist that at least one of the two following statements (both of which Berkeley affirmed) must be false:

- Extension and impenetrability exist only in the mind.
- A thing is material if and only if it is both extended and impenetrable.

Berkeley never used the phrase "external world," and he might well have said that philosophers who wished to use that term of art were free to use it in any sense that pleased them. He would, however, have vehemently rejected the thesis that either statement 1 or statement 2 must be false. His argument against that thesis would have proceeded along the following lines:

Statement 2 is true by definition, so its denial is nonsensical. As to statement 1, extension must mean either visual extension (the quality called "extension" when one is speaking of ideas presented by the sense of sight) or tactile extension (the quality called "extension" when one is speaking of ideas presented by the sense of touch). It follows that neither visual extension nor tactile extension can exist outside the mind. Similar reasoning applies to the term impenetrability. Therefore, neither extension nor impenetrability exists outside the mind; i.e., statement 1 is true.

Because Berkeley explicitly affirmed the existence of "material" objects and implicitly accepted the existence of an "external world," the problem of the reality of an external world cannot simply be identified with the question of the truth of Berkeleyan idealism (i.e., the external world is real if and only if Berkeleyan idealism is false). As indicated above, however, most philosophers regard the problem of the reality of the external world as practically equivalent to the question of the truth of Berkeleyan idealism, in part because they reject Berkeley's eccentric understandings of the terms in which the problem is typically formulated.

2.2.2 The Problem in 20th-Century and Later Philosophy

In his famous 1939 British Academy lecture, "Proof of an External World," the English philosopher G.E. Moore (1873–1958) offered a very simple "proof" of the reality of an external world. The importance of the proof, however, lies not so much in the proof itself as in certain arguments that Moore presented to establish that it was indeed a proof. The proof was essentially this: Moore displayed one of his hands to his audience and said, "Here is one hand." He then displayed his other hand and said, "and here is another." He continued by presenting a careful and very precise piece of reasoning intended to establish the conclusion that the "here-is-a-hand" argument was indeed a proof of the reality of an external world—that, considered in relation to the proposition "there is an external world," the here-is-a-hand argument satisfied all the requirements for being a proof that anyone could reasonably impose. One essential premise of the reasoning by which Moore claimed to have demonstrated that the here-is-a-hand argument was a proof of the existence of an external world was the assertion that he and everyone in his audience knew that there was a pair of hands before them.

Notwithstanding the blunt simplicity of Moore's proof—a characteristic that struck some philosophers as slightly absurd—it is undeniable that Moore regarded the problem of the reality of the external world with deep seriousness. In his view, "Is there an external world?" is a legitimate philosophical question, just as "Did the human species die out in the 13th century?" is a legitimate historical question. Most later philosophers, however, differed from Moore on that point. In various ways, they attempted to show that the question was not genuinely philosophical or that it was not even a real question but instead a form of nonsense. Those philosophers held that the historical problem of the reality of the external world was the product of fundamental misunderstandings of the nature of human knowledge, the nature of language, or even the nature of human being.

Such criticism was characteristic of logical positivism, an important school of analytic philosophy that flourished between the two World Wars. For the logical positivists, the sentences "There is an external world" and "There is not an external world" are both literally meaningless, as are many other metaphysical utterances. That position is a consequence of the logical positivists' "verifiability principle," according to which a sentence is literally meaningful if and only if it is either in principle empirically verifiable (or falsifiable) or a tautology. Because no possible experience (no possible experiment or observation) could either prove or disprove that there is an external world, all statements about its existence or nonexistence are nonsensical.

The Austrian-born philosopher Ludwig Wittgenstein (1889–1951) and the German philosopher Martin Heidegger (1889–1976), who are widely regarded as the two most important philosophers of the 20th century, believed that the question of the reality of the external world could be meaningfully posed only within philosophical traditions that were founded upon, and perpetuated, misapprehensions of some fundamental aspect of human experience or the human condition. For Wittgenstein, the misapprehensions concerned human language and thought; for Heidegger, they concerned human being.

Wittgenstein's famous "private-language argument," which appears in his posthumously published work Philosophische Untersuchungen, can be read as implying that if there were no external world, all language would be without meaning—from which it would follow that the question "Is there an external world?" is itself meaningless unless there is an external world. If Wittgenstein is correct, the question "Is there an external world?" is effectively answered in its being posed, as are "Is there such a thing as language?" and "Does anyone ever ask a question?" In a later work Über Gewissheit, Wittgenstein insisted on a radical distinction between certitude and knowledge, holding that the former is not merely a surer form of the latter. Instead, certitude is the background or setting in which the "language games" of knowing, doubting, and believing take place. Ultimately, what is certain is that which is presupposed or taken for granted in the social activities of a community.

In his early masterpiece Sein und Zeit, Heidegger characterized the philosophy of Descartes—and, by extension, all of modern epistemology—as positing a division between an "inner" world of subjective mental experiences and a hypothetical "outer" world of objective material things. The two worlds were in principle completely independent of each other, and the only possible relation between the two was that of "representation," whereby certain elements of mental experience could represent or correspond to certain features of the objective material world. The task of philosophy, according to the Cartesian conception, was to show how or to what extent the relation of representation might be veridical.

According to Heidegger, however, the two "worlds" were not at all independent; to the contrary, each was a distorted abstraction of a primordial and unified Dasein—the mode of human being—which was inherently already involved with and caught up in a world that the Cartesian tradition had misconceived as independent of isolated human subjects. Commenting on Kant's "scandal to philosophy," Heidegger wrote: "The 'scandal to philosophy' does not lie in the fact that this proof has not yet been given, but rather in the fact that such proofs are continually expected and attempted."

It should be remarked, however, that in the 20th century—Heidegger's century—such proofs were not actually continually expected or attempted. Even Moore's "proof" was best understood not as a genuine attempt to prove the existence of an external world but rather as a way of raising a pointed philosophical question about what the skeptical demand for a proof of the reality of the external world really amounted to. And that question is, "Why does a certain trivial exercise not count as a proof of the reality of an external world?"

The remarkable revival of metaphysics among analytic philosophers in the last quarter of the 20th century did nothing to reawaken interest in the question of the reality of the external world. Subsequent analytic metaphysics was concerned either with problems that had no bearing on that question (e.g., problems regarding modality, ontology, and the nature of time) or with questions about the metaphysics of the physical or material world. The sole exception to that generalization is provided by a few defenses of idealism, such as the essay "Idealism Vindicated" (2007), by the American philosopher Robert Merrihew Adams, and A World for Us: The Case for Phenomenalistic Idealism (2008), by the English philosopher John Foster.

Mind and Body

Human beings seem to have properties of two quite different kinds: physical properties, such as size and weight, and mental properties, such as feeling pain or believing that Tokyo is the capital of Japan, which imply sensation or thought. Among the properties of persons, mental properties have seemed to many philosophers to be in some sense deeper or more fundamental than physical properties. That idea is supported by a famous observation made by Descartes: that whereas it is at least conceivable that people should be mistaken about their physical properties, it is inconceivable that their mental properties should be anything other than what they seem to them to be.

There are many theories about the existence of and relation between physical and mental properties. At one extreme there are idealists, who deny the existence of physical properties, and at the other there are behaviourists and eliminative materialists, who deny the existence of mental properties. Most philosophers, however, believe that properties of both kinds exist. Philosophical theories of mind may be categorized according to the ways in which they conceive of the relation between physical and mental properties and according to their implications regarding the so-called mind-body problem, the problem of explaining how mental events arise from or interact with physical events. Historically, three types of theory have been most influential: psychophysical monism, property dualism, and psychophysical dualism.

According to psychophysical monism, the physical and mental properties of human beings are properties of the same thing: of their bodies or of parts of their bodies, such as the cerebral cortex or the nervous system. Psychophysical monists also believe that the mental properties of a thing are completely determined by its physical properties. Thus, a perfect physical duplicate of a thinking, feeling human being would, of necessity, have exactly the same mental properties as that human being. Psychophysical monists are almost all proponents of the identity theory, according to which mental events are the same as or identical to physical events.

Property dualists agree with psychophysical monists that the physical and mental properties of human beings are properties of the same things but reject the other thesis of the monists, that the physical properties of a thing necessarily determine its mental properties. They hold that it is at least metaphysically possible to assume that there are two beings with identical physical properties but different mental properties. That possibility, moreover, implies that mental properties are nonphysical properties—hence the term property dualism. The so-called double-aspect theory of Benedict de Spinoza (1632–77) and Arthur Schopenhauer (1788–1860) is probably best categorized as a form of property dualism.

According to psychophysical dualism, the physical properties of human beings are properties of their bodies and the mental properties of human beings are properties of their minds or souls—a person's mind or soul being an immaterial substance wholly distinct from the physical substance that is that person's body.

Among psychophysical dualists, dualistic interactionists hold that the body and the mind interact—that the mind causally affects the body and the body causally affects the mind. Dualistic interactionists seem to be committed to the position that the physical world is not causally closed—i.e., that physical events cannot always be completely explained by reference to earlier physical events and the laws of physics. That position, however, would seem to be inconsistent with the conservation laws that are fundamental to modern physics.

Other psychophysical dualists, known as occasionalists, have maintained that the apparent causal interaction between mind and body is only apparent: mental and physical changes are coordinated by the direct action of God. Like interactionists, however, occasionalists seem to be committed to the thesis that there are physical events that cannot be explained in terms of earlier physical events.

The theory of preestablished harmony, due to Gottfried Wilhelm Leibniz (1646–1716), in some ways resembles occasionalism but avoids the problem of inconsistency with the closure of the physical world by postulating separate physical and mental realms, each of which unfolds deterministically with the passage of time according to its own laws; the two realms do not interact but have been created (by God) in such a way that they are in perfect harmony with each other.

The very unclear position called epiphenomenalism is sometimes categorized as a form of psychophysical dualism according to which the body affects the mind but the mind does not affect the body. Thus, when a human being wills a certain bodily movement and that movement occurs, the movement is caused entirely by prior physical states of the body. The corresponding act of will, however, is also caused by prior physical states. As a result, the act of will seems to its subject to be the cause of the movement. It is, however, probably better to think of epiphenomenalism not as a form of psychophysical dualism but as a form of property dualism according to which both mental events and physical events are entirely caused by physical events. Partly because few philosophers have thought of themselves as epiphenomenalists, it is difficult to categorize that view under any familiar type of philosophical theory of mind. The best-known modern epiphenomenalist was the biologist Thomas Henry Huxley (1825–95).

Existence

Although metaphysicians have had a great deal to say about the existence of various things, they have had less to say about existence itself—about the content of the concept of existence or about the meaning of the word existence. They have said enough, however, to make possible a taxonomy of theories of existence. Such a taxonomy can be presented as a list of pairs of opposed or contradictory theses about the nature of existence.

- Some metaphysicians have affirmed, and others have denied, that existence is the same as being. It might seem obvious that "Mountains higher than Mont Blanc exist" and "There are mountains higher than Mont Blanc" are two ways of saying the same thing. But some metaphysicians believe that there are things that do not exist—fictional characters, for example, or the Greek gods. Their position is that, although such things certainly do not exist, the fact that there are such things implies that they have being. If something can "be" without existing, they argue, then existence and being must be distinguished.

- Some metaphysicians have affirmed, and others have denied, that existence is a barren or empty or trivial concept. G.W.F. Hegel (1770–1831), for example, referred to being as "the very poorest and most abstract" of all categories.
- Some metaphysicians have affirmed, and others have denied, that the word exist means the same thing in all its applications. For example, mathematicians habitually speak of the "existence" of abstract, mathematical objects like numbers or functions. Metaphysicians, as well as philosophers of mathematics, differ on the question of whether existence in such assertions means the same as it does when it is applied to persons and other tangible, visible things.
- Some metaphysicians have affirmed, and others have denied, that the being (or existence) of one object may be "more perfect," or "of a higher degree," than the being (or existence) of another. A classic expression of that idea is the analogy of the divided line, representing a fourfold hierarchy of being, from the Republic of Plato (c. 428–c. 348 BCE): images and shadows participate in being very imperfectly, sensible objects less imperfectly, and "mathematicals" less imperfectly still. But the eternal, unchangeable forms—and only they—exhibit being perfectly.
- Some metaphysicians have affirmed, and others have denied, that existence is a property or attribute of everything that exists. Kant, the most famous critic of the thesis, identified it as the fallacy on which the ontological argument for the existence of God depends. Deniers of the thesis have maintained that "existence" statements are only apparently about the things that are their grammatical subjects and so cannot be understood as attributing a certain property to those things. Gottlob Frege (1848–1925), for example, held that the statement "Horses exist" really means "The number of objects that fall under the concept horse is not zero."

A theory of existence may be identified with some combination of the theses discussed above. It should be noted, however, that some combinations are inconsistent, or at least apparently so. For example, anyone who accepts Frege's account of existence seems to be committed to the theses that existence is a trivial concept, that there is no distinction to be made between being and existence, that existence means the same thing in all its applications, and that existence is not something that one thing can exhibit more perfectly or in some higher degree than another.

2.2.3 Universals and Particulars

Many philosophers have believed that, in addition to particular things, there are "general" things of which particular things are instances or examples or cases. They have believed, for example, that, in addition to particular horses, the world contains the species Equus caballus, a general thing of which every horse is an instance. The Latin word for such general things is universalia.

The complex history of the Latin term may be briefly summarized as follows.

Plato had used the adverbial phrase kath' holou ("as a whole") in statements such as, "I am interested not in this or that case of virtue but in virtue taken kath' holou." His student Aristotle (384–322 BCE) used the word katholou, a noun coined from the adverbial phrase, as a name for those things that could be predicated or said of a thing—thus, "being virtuous" and "being white" are katholou. Later philosophers writing in Latin, seeking a noun corresponding to katholou, settled on universale and universalia, the neuter singular and neuter plural forms, respectively, of the adjective universalis ("universal"). They chose universalis for that purpose because it is derived from another adjective, universus, which means "taken as a whole."

A strikingly high proportion of the writings of medieval philosophers is directed at disputes about the nature of universalia and the nature of their relation to the "particulars" that are their instances. The medieval interest in universals is at least partly to be explained by the respect for philosophical authority characteristic of the Middle Ages and by the fact that the two greatest authorities of antiquity, Plato and Aristotle, had disagreed about such matters.

According to Plato, when the same general word (e.g., horse, spear, river) applies to different particulars, it does so by virtue of the fact that those things bear a common relation to a certain form or idea—a supersensible, eternal, and changeless being. If, for example, the word horse applies to each of two particulars, it is only because both of them fall under or "participate" in the form Horse. If the application to particular things of general terms like horse were not in some way guided by an understanding of their associated forms, the fact that speakers apply the word to each of two particular things would be a mere accident and would not reflect any common nature among the things so designated—as is the case with the application of the name Heraclitus to each of a dozen or so ancient Greeks. It seems to be an indisputable consequence of Plato's theory of forms that the existence of any form does not require that there be a particular thing that participates in it. If, for example, horses were to become extinct, the form Horse, being eternal and changeless, would continue to exist.

Aristotle agreed with Plato on one important point: that when a general term is applied to many particular things, its application is "guided" by knowledge of something that those things have in common. Aristotle denied, however, that what (for example) all horses have in common is participation in a changeless, eternal, independently existing form. In his view, what all horses have in common is something that inheres in each horse—in each of the multiplicity of particular horses.

Four schools of thought emerged from the medieval disputes about universals and their relation to particulars.

- Platonists, or Platonic realists, affirmed the existence of universalia ante rem (or ante res): universals "before the thing" (or "before things"). That is, they held that general things—the things that account for the fact that a general word applies to many particulars—exist independently of those particulars. According to medieval Platonic realism, before God created any horses, the species horse, or the attribute "being a horse" (or both), already existed. Some Platonists, concerned to avoid the implication that anything might exist independently of God, identified universalia ante rem with ideas in the mind of God.
- Aristotelians, or Aristotelian realists, affirmed the existence of universals but contended that they were universalia in re (or in rebus): universals "in the thing" (or "in things"). According to Aristotelians, universals exist only as constituents of particulars. The word horse applies to two things by virtue of their having a certain universal as a common constituent—the species horse or the attribute "being a horse" (or both). If God had not been pleased to create horses, the Aristotelians maintained, neither the species horse nor the attribute "being a horse" would have existed.
- Nominalists denied the existence of both universalia ante rem and universalia in re. Reality, they maintained, consists entirely of particulars. The word horse, then, does not refer to a universal, whether before or in things, but simply denotes all horses. Some nominalists did not so much deny the existence of universals as identify them with the general terms whose application Platonists and Aristotelians had invoked "universals" to explain. A universal, such nominalists contended, is a mere "puff of sound".
- Finally, conceptualists held that universals are mental entities, confined to individual human minds. Thus, any person's use of the general term horse is governed by a concept that exists only in that person's mind. Another person's use of the word is governed by a numerically distinct concept in that other person's mind—and those two concepts may well differ in content, with the consequence that the two speakers apply horse to different things. There is no question of a speaker's associating the "wrong" concept with the word horse—except in so far as that concept leads the speaker to apply the word to objects different from those to which the majority of fellow speakers apply it.

The schools described above are, to a very large extent, abstractions. In practice, it is often difficult to place a given philosopher in any one of them to the complete exclusion of the others. As a very general rule, it can be said that philosophers writing in Latin in late antiquity tended to be Platonic realists, that philosophers of the high Middle Ages tended to be Aristotelian realists, and that philosophers of the late Middle Ages tended to be nominalists.

Theories of universals received relatively little attention from philosophers of the Renaissance and modern periods. That was particularly true of the empirically minded British philosophers of the 18th century, who generally adopted an uncritical nominalism and who tended to regard any attempt at serious discussion of the nature of universals as so much Scholastic quibbling.

In the 20th century there was a remarkable resurgence of interest in theories of universals. There were, on the one hand, a significant number of philosophers who defended the thesis that an appeal to Aristotelian universals explains various features of the world (e.g., that a green book and a green apple have something in common). And there was, on the other hand, a revival of Platonic realism, in large part a consequence of the realism about mathematical objects advocated by Bertrand Russell (1872–1970) and the philosophy of logic of W.V.O. Quine (1908–2000), particularly his views on quantification and "ontological commitment."

Causation

The word causation applies to relations of two distinct kinds. To resolve the ambiguity, it is necessary to distinguish between the idea of "the" causal relation and the idea of "a" causal relation. The terms of "the" causal relation are causes and effects, and causes and effects are, by definition, events or facts or states of affairs. That relation—the relation between cause and effect—is expressed by phrases such as caused, was the cause of, was one of the causes of, led to, and had as an effect. Its converse is expressed by phrases such as was caused by, was an effect of, was due to, occurred because, was the result of, and was a consequence of. To affirm that an event or fact or state of affairs (a cause) bears the causal relation to another such item (an effect) is to imply that the former in some sense explains the latter.

It will be convenient to have an unambiguous name for the causal relation. In the remainder of the present section, this relation will be called the cause-effect relation.

By contrast, the terms of "a" causal relation—there are many of them—are substances or concrete objects or particular things or entities. That is, they are items such as fingers and billiard balls and souls—as opposed to items such as a finger's pressing a button, a billiard ball's striking another billiard ball, and a soul's performing an act of will. Such causal relations are expressed by words and phrases such as pushed, pulled, touched, bent, struck, cut, kissed, killed, bored a hole in, exerted a force on, acted on, affected, and brought into existence.

Again, it will be convenient to have an unambiguous name for such causal relations. In the remainder of the present section, they will be called agent-patient relations, a term that indicates that they are relations that a thing bears to another thing by virtue of the way in which the former acts on the latter.

The philosophical topic of causation comprises the cause-effect relation, the many agent-patient relations, and the relation between the two, for the cause-effect relation and the many agent-patient relations are intimately connected. Agent-patient relations, for example, often figure in descriptions of causes, as in the statement, "She died as the result of being struck by a car." That statement asserts that the event or state of affairs "(her) being struck by a car" bore the cause-effect relation to the event or state of affairs "(her having) died," and the former item is described in terms of an agent-patient relation ("being struck") between a certain vehicle and a certain human being.

In Greek and medieval philosophical texts, the Greek words aition (or aitia) and the Latin word causa—each of which is usually translated as "cause"—almost always refer to agent-patient relations of one kind or another. Present-day philosophical treatments of causation, however, are devoted almost entirely to the cause-effect relation, reflecting the enormously influential work of the Scottish Enlightenment philosopher David Hume (1711–76) but also in part the influence on philosophy of Newtonian and post-Newtonian science. Although not all philosophers (even within the Anglophone philosophical tradition) accept Hume's account of causation, most would agree with Hume on one important point: the central topic in a philosophical study of causation should be the cause-effect relation. Most philosophers would, in fact, suppose that agent-patient relations are not an important part of the topic of causation and that they should in any case be understood or analyzed in terms of the cause-effect relation. For example, they would for the most part subscribe to the following thesis:

A satisfactory philosophical analysis of the statement "The Earth exerts a gravitational force on the Moon" should take the form of a statement to the effect that some (specified) event or fact or state of affairs that involves the Earth is the cause of a certain effect—to wit, some (specified) event or fact or state of affairs that involves the Moon.

According to Hume, to say that event A is the cause of event B is to say that events that are very much like A have always been followed by events that are very much like B—that events very much like A have been "constantly conjoined" in human experience with events very much like B. For example, to say that a billiard ball's being struck by another billiard ball was the cause of the movement of the former ball is to say that the first ball was stuck by the second and that events very much like that striking have always been followed by events very much like the first ball's movement. Hume thus rejects the idea of a "necessary connection" between cause and effect: the concept of one ball's coming into contact with another does not contain or imply the concept of the latter's moving. The idea of such a connection, he contends, arises from the experience of constant conjunction, which creates a disposition in the human mind to expect an event B after having observed an event A, given that events very much like A have always been followed by events very much like B.

Many later philosophers in the Anglophone philosophical tradition have accepted some refinement of Hume's theory of causation. (And refinements are obviously necessary: if people always hear a church bell ringing immediately after the chiming of their own clocks, it does not follow that the clocks' chiming is the cause of the church bell's ringing.) Contemporary Humean theories are usually framed in terms of the laws of nature. One such formulation, for example, is:

An event A is the cause of an event B if and only if there is some property F of A and some property G of B such that 'an event with property F is always followed by an event with property G' is a law of nature or a logical consequence of such a law.

They maintain, moreover, that laws of nature are merely the most general of the observed regularities in nature and are in no sense necessary truths.

Although he admired Hume's analysis of causation, Kant maintained that what he called "causality" did indeed include the idea of a necessary connection between cause and effect. In the Critique of Pure Reason, he argued that causality is an a priori category, or pure concept of the understanding. Because it does not derive from experience, it is possible to have the concept of causation without directly perceiving the necessity that is inherent in it. Because the categories apply only to experience, however, they cannot be used to establish the existence or nature of anything that is not itself an object of possible experience (e.g., God, freedom, and immortality). Thus, the traditional cosmological argument, which attempts to prove the existence of a "first cause" of the universe (i.e., God), is invalid, according to Kant.

Few present-day philosophers would agree with Kant that causation is an a priori category. Indeed, most contemporary philosophers of causation either accept some version of Hume's account or at least share his assumption that the central topic in the philosophy of causation should be the analysis of the cause-effect relation. There is, however, one important exception to that generalization. Some 20th-century writers on free will revived the medieval idea of substantial causation, which can be understood as a hybrid of an agent-patient relation and the cause-effect relation. Those philosophers contended that it is possible for an event (or fact or state of affairs) to have been caused not by an earlier event (or fact or state of affairs) but by a substance. It may be, they argued, that, when people raise their arms, the people themselves—and not any event or fact or state of affairs—are the cause of their arms rising. Indeed, they maintained, that is what a free act consists of: an act caused by its agent (a substance) and not caused by any prior event or fact or state of affairs.

Substance

Although substance is one of the most important ideas in metaphysics, philosophers disagree about which entities are substances. For Aristotle, the first philosopher to make substance a central concept in his thought, the best examples of substances (among tangible, visible things) were living organisms. For Benedict de Spinoza (1632–77), who also gave the concept a central place in his philosophy, there was only one substance, which constitutes the whole of reality. Spinoza held that living organisms are mere "finite modes" of the one (infinite) substance.

If one assumes that Aristotle and Spinoza employed the same concept of substance (Aristotle holding that the best examples of substances were biological organisms, and Spinoza taking the view that the only substance was the whole of reality), then the concept must be very abstract indeed. A suitably broad concept might be set out as follows: a substance is a particular that exists "in its own right"—i.e., a particular thing that could exist independently of other particular things. But that attempt at explaining the concept of substance raises the following questions: What is the concept of substance opposed to? What sort of particular is not a substance? In other words, what particulars might be said to be incapable of existing independently of other particulars?

Such questions are best answered by giving examples. Some things are present only "in" other things: e.g., a smile, a wrinkle, a surface, a hole, a reflection, or a shadow. There is a clear sense in which such items, even if one is willing to grant real existence to them, do not exist in their own right. They might be called "ontological parasites," things incapable of existing apart from the things that are their "hosts." If one supposed that a carpet could, in metaphysical theory if not in physical fact, exist apart from all other things, one would be supposing that the carpet was a substance, but no one would suppose that a wrinkle in that carpet could be a substance. The carpet may or may not exist in its own right, but the wrinkle certainly does not.

Aristotle had called things that exist in their own right prōtai ousiai ("primary beings"; singular prōtē ousia), which make up the most important of his ontological categories. Several features define prōtai ousiai: they are subjects of predication that cannot themselves be predicated of things (they are not universals); things exist "in" them, but they do not exist "in" things (they are not "accidents," like Socrates' wisdom or his ironic smile); and they have determinate identities (essences). The last feature could be expressed in modern terms as follows: if a prōtē ousia x exists at a certain time and a prōtē ousia y exists at some other time, there is a determinate answer to the question of whether x and y are one and the same thing, or numerically identical; and the question of whether a given prōtē ousia would exist in some specific set of counterfactual circumstances must likewise have a determinate answer. It is difficult to suppose that smiles or wrinkles or holes have this sort of determinate identity. To ask whether the smile that Socrates smiled today is the same as the smile that he smiled yesterday can be understood only as a question about descriptive identity, the relation between two distinct things whereby they are exactly like each other in every respect.

Aristotle used ousia not only as a count noun but also as a mass term. For example, he asked not only questions like "Is Socrates a ousia?" and "What is a ousia?" but also questions like "What is the ousia of Socrates?" and "What is ousia?" In the count-noun sense of the term, Aristotle identified at least some ousia with ta hupokeimena ("underlying things"; singular to hupokeimenon). Socrates, for example, is a to hupokeimenon in that he "lies under" the in rebus ("in the things") universals under which he falls and the accidents that inhere in him. To hupokeimenon has an approximate Latin equivalent in substantia, "that which stands under." Owing both to the close association of ousia and to hupokeimena in Aristotle's philosophy and to the absence of a suitable Latin equivalent of ousia (the closest analogue, essentia, a made-up Latin word formed in imitation of ousia, was used for another purpose), substantia became the customary Latin translation of the count noun ousia. A substantia or substance is thus a particular that is capable of "standing on its own." A substance may indeed depend on the action of other substances for its existence: it may have been brought into existence by the prior operations of other substances, and it may depend on the concurrent operations of other substances to continue in existence. But it does not depend on other things for its existence in the manner in which a wrinkle or a hole in a carpet depends on the carpet for its existence.

Although there is no universally accepted and precise definition of "substance", most philosophers would agree that certain kinds of things are not substances. For example, most philosophers who are willing to use the word at all would deny that any of the following are substances:

- Universals and other abstract objects. (It should be noted that Aristotle criticized Plato for supposing that prōtai ousiai are ante res, or "before things," universals.)
- Events, processes, or changes. Some philosophers have held that there are substances that are nontemporal, or outside time. But substances that are temporal are said to last or endure or to exist at various times. Events or processes, on the other hand, are said to happen, occur, or take place.
- Stuffs, such as flesh, iron, or butter. Although a common meaning of "substance" is stuff or matter, Aristotle criticized earlier philosophers (specifically, the pre-Socratic cosmologists) for supposing that a prōtē ousia could be a stuff such as water, air, fire, or earth.

The question of whether there are substances continues to be one of the central problems of metaphysics. Several closely related questions are the following. How, precisely, should the concept of substance be understood? Among the sorts of things that human beings frequently encounter, which (if any) are substances? If there are substances, how many of them are there? What kinds of substances are there?

2.3 IDENTITY

When one says, for example, that the room in which Hegel lectured was identical with the room in which Schopenhauer lectured, there are two quite different things that one might mean. The first is that the two philosophers lectured in rooms that were in different places but were of the same dimensions and were in every other respect exact duplicates of each other. (It is in this sense of "identical" that monozygotic twins are said to be identical.) The second is that Hegel and Schopenhauer lectured in one and the same room (though presumably at different hours). Identity of the former sort is called descriptive identity, and identity of the latter sort is called numerical identity—"numerical" because, if x and y are identical in that sense, there is only one of them; some one thing is both x and y. Although the concept of descriptive identity has received a considerable amount of attention from philosophers, numerical identity is the more important of the two concepts for metaphysics.

The logical properties of numerical identity have been precisely codified by logicians, who express it by the sign "=." The sign has been borrowed from mathematics, but (the logicians insist) without any change of meaning. According to logicians, a mathematical equation—a formula that consists of two expressions surrounding the symbol "="—is simply a statement of numerical identity. The equation $7 + 5 = 2 \times 6$, for example, differs from the statement "Mark Twain is (numerically) identical with Samuel Clemens" in its subject matter—the latter is about a person (the person who was called both "Mark Twain" and "Samuel Clemens") and the former about a number (the number that is designated by both "7 + 5" and "2 × 6")—but not in its logical structure. The properties of "=" are, according to the standard formal logic of identity, exactly those expressed by two axioms: $x = x$, which says that any object x is identical with x—that is, with itself—and $(x = y)\ (Fx = Fy)$, which says that if an object x and an object y are identical, then something F is true of x if and only if F is also true of y. Thus, because Mark Twain and Samuel Clemens were identical, Mark Twain was fond of buttered toast if and only if Samuel Clemens was fond of buttered toast. The latter axiom has been called both the principle of the indiscernibility of identicals and Leibniz's law (see identity of indiscernibles). It can be intuitively stated as follows: if x is identical with y, whatever is true of x is true of y and whatever is true of y is true of x.

There are apparent exceptions to Leibniz's law. Consider, for example, the following argument:

- Mark Twain chose that name as a nom de plume.
- Mark Twain was identical with Samuel Clemens.
- Therefore, Samuel Clemens chose that name as a nom de plume.

It might appear that Leibniz's law incorrectly implies that the preceding inference is valid. Almost all philosophers agree, however, that the argument is not a counterexample to Leibinz's law, because the phrase "chose that name as a nom de plume" does not really express something that can be true of or false of someone.

The following argument, often attributed to Descartes, is widely regarded by philosophers as a similarly fallacious attempt to apply Leibinz's law.

- The following is true of my body: I can imagine that it does not really exist, though it seems to me that it does exist. (For example, I can imagine that I have been dreaming my whole life through and that the world of material things that I seem to perceive around me is no more than a figment of my long dream.)
- The following is not true of me: I can imagine that I do not really exist, though it seems to me that I do exist.
- Therefore, I am not identical with my body.

The argument is a "contrapositive" application of Leibniz's law. The law implies that if a person and that person's body are identical, then what is true of either is true of the other; it follows that if something is true of a person's body that is not true of that person, then the person and the person's body are not identical.

The standard criticism of this argument is that the phrase "I can imagine that x does not exist, though it seems to me that it does exist" does not express something that can be true or false of a thing. A moment's reflection shows that if those words did in fact express something that could be true or false of a thing, then no first-person identity statement more informative than "I am I" or "I am myself" could be true. If, for example, Lee Harvey Oswald had been brought to trial for having murdered John F. Kennedy, he could have established his innocence by arguing as follows:

- The following is true of the murderer of John F. Kennedy: I can imagine that he does not exist, though it seems to me that he does exist.
- The following is not true of me: I can imagine that I do not really exist, though it seems to me that I do exist.
- Therefore, I am not identical with the murderer of John F. Kennedy.

It is easy to prove that the two axioms of identity (Lebiniz's law and "Everything is identical with itself") logically imply that identity has the following features: it is symmetrical (if Mark Twain is identical with Samuel Clemens, then Samuel Clemens is identical with Mark Twain); it is transitive (if Byzantium is identical with Constantinople and if Constantinople is identical with Istanbul, then Byzantium is identical with Istanbul); and it conforms to "Euclid's law," or the principle that identicals may be substituted for identicals (if angle A is twice as large as angle B and if angle C is identical with angle A, then angle C is twice as large as B). Indeed, Leibniz's law is nothing more than a somewhat more careful statement of Euclid's law.

The principle of the indiscernibility of identicals must be carefully distinguished from its contrapositive, the principle of the identity of indiscernibles. The latter principle may be stated as follows: "If whatever is true of x is also true of y and if whatever is true of y is also true of x, then x and y are identical." The fact that the principle of the indiscernibility of identicals is also called Leibniz's law and the fact that the principle of the identity of indiscernibles plays a central role in Leibniz's metaphysics have no doubt encouraged confusion between the two principles.

The principle of the identity of indiscernibles is a trivial truth if there are "individual essences"—that is, properties of a thing that consist of its being that particular thing and no other thing. If there are individual essences, then the principle would imply that each thing is identical with itself and with no other thing. However, many philosophers doubt that such individual essences really exist, and almost all philosophers who have expressed an opinion on the question believe that, individual essences apart, the principle of the identity of indiscernibles is not a necessary truth; that is, it is possible to imagine without contradiction a universe in which the principle would be false.

Some of the most important philosophical debates about identity have to do with identity across time, particularly the identity of persons across time. The thesis that there is such a thing as identity across time is simply the view that one and the same entity may exist at more than one time—or, equivalently, that it is possible for a thing existing at one time and a thing existing at another time to be numerically identical. It would seem that almost everyone unreflectively believes that there are real cases of identity across time. Any history of physics, for example, will state that a certain person, Albert Einstein, formulated the special theory of relativity in 1905 and formulated the general theory of relativity in 1915. If that statement is true, then the person who formulated the special theory of relativity in 1905 was identical with the person who formulated the general theory of relativity in 1915. Nevertheless, the commonsense assumption that Einstein in 1905 was identical with Einstein in 1915 is at least apparently inconsistent with Leibniz's law, since Einstein in 1905 and Einstein in 1915 did not have all of the same properties.

Philosophers have proposed various solutions to the preceding problem. Some would say that Einstein existed at different times in virtue of having "temporal parts" that individually occupied various points in, or segments of, time. One temporal part of Einstein, some would say, formulated the special theory, and another part formulated the general theory. Other philosophers would say that there is no problem to be solved by an appeal to temporal parts: the problem, the apparent violation of Leibniz's law, is due to a failure properly to understand what is asserted by sentences such as "The person who formulated the special theory of relativity in 1905 was 26 years old." What that sentence "really" says, they contend, is that the person who formulated the special theory of relativity in 1905 was 26 years old when he formulated the special theory of relativity. When that fact is appreciated, they go on to say, the apparent violation of Leibniz's law vanishes, for the person who formulated the general theory of relativity in 1915 also had that property—that is, the property of being 26 years old when he formulated the special theory of relativity.

One factor that makes problems of personal identity particularly difficult is the tension between the psychological and physical aspects of common intuitions about what it is for the same person to exist at different times. If, for example, a person's memory is entirely obliterated by some procedure that leaves the person's body unaffected, does that person still exist? Or, if a science-fictional "transporter" or "teleportation machine" should become a reality, would the human being who emerged from teleportation by such a machine be the same person as the human being who had entered the machine a spilt second earlier? Possible solutions vary with the concept of identity one employs and the metaphysics of parts and wholes one appeals to, but any plausible solution must be consistent with Leibniz's law.

Many of the most challenging problems about identity across time are raised by cases that involve a thing's changing its parts. An ancient example, known as "the ship of Theseus," may be posed as follows. A new ship, made entirely of wooden planks, is named the Ariadne. The Ariadne puts to sea, and, while it is sailing, the planks of which it is constructed are replaced by new planks, each replacement plank being descriptively identical with the plank it replaces. The original planks are taken ashore and stored in Piraeus. After all of the planks have been replaced, the ship constructed entirely of the replacement planks is still sailing in the Aegean Sea. The old planks are then assembled in a dry dock in Piraeus to form a new ship. The planks that constitute the Piraean ship stand in the same spatial relations to one another as they did when they first constituted the Ariadne. The Aegean ship and the Piraean ship are obviously not the same ship, since they are in different places at the same time. But which is the Ariadne? The problem of the ship of Theseus is the problem of finding the right answer to that question.

One might argue that the Aegean ship is the Ariadne, because a ship does not cease to exist when one of its constituent planks is replaced; hence, during the gradual replacement of its planks, there was no point at which the Ariadne ceased to be the ship it originally was. But one could also argue that the Piraean ship is the Ariadne, because the Piraean ship and the Ariadne are composed of exactly the same planks arranged in exactly the same way.

Again, possible solutions to the problem will vary depending on concept of identity and on the metaphysics of parts and wholes, but any solution must be consistent with Leibniz's law.

The concept of numerical identity has also figured essentially in philosophical critiques of various Christian theological doctrines, particularly those of the Trinity, the Incarnation, and the Eucharist. Many philosophers have held that, for example, the doctrine of the Trinity violates the principle of the transitivity of identity, since it implies, for example, that the Father and the Son are identical with God but not identical with each other.

The English Roman Catholic philosopher Peter Geach (1916–2013) proposed a radical solution to the theological problem regarding the transitivity of identity. According to Geach, there is no such thing as numerical identity; there are, instead, many relations of the form "is the same F as," where "F" is a sortal term designating a kind of thing (e.g., "human being," "animal," "living organism"; "plank," "ship," "material object"; and so on). Geach maintained that no rule of logic licenses an inference from "x is the same F as y" to "x is the same G as y," if "F" and "G" represent logically independent sortal terms. Accordingly, as far as logic is concerned, it is perfectly possible for there to be entities x and y such that: (1) x is the same F as y, but (2) x is not the same G as y. Geach's theory would thus permit one to reformulate the Trinitarian implication above as follows: (1) the Father is the same God as the Son (i.e., the Father and the Son are both God), but (2) the Father is not the same person as the Son.

Geach's theory is characterized as the view that "identity is relative to a sortal term," or simply as the "theory of relative identity." It has attracted some attention among philosophers and logicians, but not as much as it might have had it been clear that the theory had some application outside Christian theology, a matter of some dispute. It does seem, however, that the theory of relative identity might be applied to some of the problems of identity over time discussed above. In the case of the ship of Theseus, for example, one might propose the following: (1) since there is no such relation as numerical identity, the question of whether the Ariadne is the Aegean ship or the Piraean ship is meaningless; (2) the Ariadne, the Aegean ship, and the Piraean ship are all ships and all material things; (3) the Ariadne and the Aegean ship are the same ship but not the same material thing; and (4) the Ariadne and the Piraean ship are the same material thing but not the same ship.

2.3.1 Persistence Through Time

A thing is said to persist through time, or simply to persist, if it exists at more than one moment of time. Because persistence implies that the same object exists at more than one moment of time, persistence is often referred to as identity through (or across) time. Some philosophical problems concerning persistence or identity through time have more to do with identity than with time, and others have more to do with time than with identity. Problems of the former kind were discussed in the preceding section; the present section concerns problems of the latter kind.

Theories or accounts of persistence may be divided into two broad types: those based on the view that time is very much like space and those based on the view that time and space are fundamentally different.

Metaphysicians who belong to the former school regard identity across time as closely analogous to "identity across space." The latter kind of identity is exemplified by the identity of a river that is encountered in one place with a river that is encountered in another place. What is it to say that two bridges span the same river despite the fact that they are distant from each other? The obvious answer is that the river consists of many parts or segments, and one of the bridges spans one of the segments that compose the river and the other bridge spans another of those segments. Accordingly, philosophers who regard identity across time as closely analogous to identity across space adopt a similar analysis—appealing to temporally distinct parts or segments—when analyzing problems of the former kind.

For example, suppose that a certain cat existed at the moment of its conception in 1832, at the moment of its death in 1844, at all moments in between those two moments, and at no other moments. Just as a river is composed of many spatial river-segments, a cat is composed of many temporal cat-segments, or "temporal parts" of the cat. There are, for example, the 1837-part of the cat and the 1840-part of the cat, each with a temporal extension of one year, and the June 11, 1835-part of the cat, with a temporal extension of 24 hours. To say that the kitten that Princess Victoria stroked in 1832 was identical with the dying cat that Queen Victoria stroked in 1844 is to say that Princess Victoria stroked a small, kittenish temporal part of a cat in 1832, that Queen Victoria stroked a large, moribund temporal part of a cat in 1844, and that those two temporal cat-parts were parts of the same temporal whole. That temporal whole, or whole cat, is a thing extended in time, a "space-time worm" whose endpoints are a moment in 1832 and a moment in 1844.

Thus, to persist through time is simply to have temporal parts that exist at different times. The American philosopher David Lewis (1941–2001) proposed that an object that exists at more than one time by being composed of a plurality of temporal parts should be said to "perdure," and he called the thesis that there are persisting things—and that they persist by perduring—the perdurance theory, or "perdurantism." Most Anglophone philosophers have adopted those terms.

The perdurance theory is not the only theory that treats time as being very much like space, though it is the most widely accepted theory of that kind. A rival theory maintains that, although the perdurance theory presents a correct account of what persistence would consist of if there were such a thing, there in fact is no such thing as persistence. There exists nothing in addition to the momentary things that the perdurance theory wrongly describes as temporal parts of persisting or temporally extended things. The members of the temporal sequences of momentary things that, according to the perdurance theory, compose temporally extended things are connected by a relation of causal continuity called "gen-identity," but that fact does not imply the existence of temporally extended wholes that have them as parts.

Because the gen-identity theory denies that persistence exists, it is not, strictly speaking, a theory of persistence. It is perhaps better described as a theory of the illusion of persistence, an illusion that is due to a faulty inference whereby observers who encounter a temporal succession of momentary things related by gen-identity mistakenly conclude that those momentary things are parts of a temporally extended whole.

Metaphysicians who accept the reality of persistence but regard time as fundamentally different from space deny that persisting things are, as the perdurantists hold, composed of temporal parts. The idea of a temporal part, they argue, depends essentially on a belief in the fundamental similarity of time and space, for the idea of a temporal part makes sense only if things are extended in time, and things can be supposed to be extended in time only if time is sufficiently like space that the idea of extension—an idea derived from the experience of physical distances between points in space—can, by a species of analogical generalization, be applied to the intervals between moments of time.

If such opponents of the perdurance theory are challenged to produce their own account of persistence, they will typically respond that no such account is possible. And that is because a demand for an account of persistence can be understood only as a demand for an analysis of the concept of persistence in terms of simpler or more fundamental or better-understood concepts—and, according to them, no concept is simpler or more fundamental or better understood than the concept of an object's existing at more than one time. Lewis called their position "endurantism," a term that he intended to be used in opposition to "perdurantism," and most metaphysicians have accepted the pair of terms as the standard way of referring to the two theories.

It must be said, however, that many perdurantists would object to any reference to endurantism as a "theory." Perdurantists commonly charge that, although endurantism is obviously a thesis about persistence, it is not a theory of persistence. They contend that the entire content of endurantism is contained in the statement "persisting objects are not composed of temporal parts," which is equivalent to the statement that a certain theory of persistence—their theory, perdurantism—is wrong. And, they point out, a statement that a particular theory of something is wrong, even if true, is not itself a theory of that something.

In response to such objections, some endurantists have granted that endurantism is not a theory of persistence but have insisted that it is no worse for not being one. Endurantism, they hold, is a correct philosophical thesis about persistence, and that is all that it needs to be.

Other endurantists, however, have replied to the perdurantists by denying that the entire content of endurantism is contained in the statement that perdurantism is wrong. Endurantism, they hold, is also the positive thesis—in fact, the "theory"—that, if a persisting object exists at a certain moment, then the whole of that object, rather than merely a temporal part of it, exists at that moment. They therefore maintain that the persistence of an object through time consists in its being wholly existent or present at different moments. And that thesis, they contend, is no less a theory of persistence than perdurantism is.

Many perdurantists, however, have expressed uncertainty about how such a theory should be understood. What, they wonder, does a sentence like "Victoria's cat was wholly present at every moment of its existence" mean? They point out that, if the sentence means only that Victoria's cat was a persisting thing but not a temporally extended thing composed of briefer temporal parts, then the following two theses are two statements of the same thesis, two ways of saying the same thing.

- The persistence of an object through time consists in its being wholly existent or present at different moments.
- The persistence of an object through time does not consist in its being composed of temporal parts that exist at different moments.

And, if that is so, the former statement is a negative thesis in an affirmative disguise—a thesis about what persistence is not that has been made to look like a statement about what persistence is by verbal sleight-of-hand.

If such challenges to the concept of an object's being wholly existent or present at a moment cannot be met, then perdurantism and endurantism should not be regarded as rival accounts of persistence. It should rather be conceded that perdurantism is the sole theory of persistence. Endurantism, moreover, should be described as a metaphysical thesis or position—and not as a metaphysical theory—that comprises the following two merely negative theses:

- Perdurantism is to be rejected, owing to the fact that it implies that persisting objects have temporal parts when, in fact, there are no such things as temporal parts. The concept of a temporal part is a product of a false picture of time, a picture that represents time as being much more like space than it really is.
- There exist no concepts in terms of which a theory of persistence could be stated, for any such concept would have to be simpler or more fundamental or better understood than the concept of a persisting thing, and no concept is simpler or more fundamental or better understood than the concept of a persisting thing.

2.3.2 Modality

The two statements "Tokyo is the capital of Japan" and "There is no largest prime number" are both true, but they differ in their relation to truth in the following important respect: whereas the former statement could have been false, the latter statement could not have been false. That is to say, the former statement is contingently true, or "contingent," and the latter statement is necessarily true, or "necessary."

The two statements "Yokohama is the capital of Japan" and "There is no smallest prime number" are both false. The former, however, could have been true, but the latter could not have been true. That is to say, the former statement is contingently false (it is false but possibly true, or "possible"), and the latter statement is necessarily false, or "impossible."

The concepts expressed by the words within quotation marks in the preceding two paragraphs are called "modal" concepts. The term modal originated in a medieval theory of necessity and contingency according to which statements can be true in various "modes." For example, the mode in which "Tokyo is the capital of Japan" is true is contingency, while the mode in which "There is no largest prime number" is true is necessity. Although philosophers no longer speak of contingency and necessity as modes of truth, the term modal—meaning "pertaining to possibility, impossibility, necessity, and contingency"—has been retained. Similarly, the noun modality is simply an abbreviated way of referring to any or all of the categories of possibility, impossibility, necessity, and contingency.

Modality that has to do with the truth or falsity of propositions is sometimes called "alethic" modality. Alethic modality is contrasted with "epistemic" modality, which is the kind of modality expressed by phrases such as "For all anyone knows...," "No one knows whether...," "It must be the case that...," "It couldn't be the case that...," and so on. A relatively oblique expression of epistemic modality, for example, is the statement "It is possible that this number is prime," as uttered by a mathematician. Although the number in question is either prime or not prime, and the number could not have been not prime if it is prime or prime if it is not prime, the mathematician's statement means only that no known mathematical consideration lends any support either to the thesis that the number is prime or to the thesis that that the number is not prime.

In addition to the modal adjectives "contingent," "possible," "necessary," and "impossible," the vocabulary of modality includes the modal operators "it is possible that" and "it is necessary that." The following two schemas display the relation between the operators and their corresponding adjectives (the term "p" represents any proposition): (1) it is possible that p if and only if p is possible, and (2) it is necessary that p if and only p is necessary.

All modal terms can be given definitions that involve only one "basic" or "primitive" modal term, whether possible or necessary. For example, one might start with necessary and define possible as follows: "It is possible that p if and only if it is not the case that it is necessary that it is not the case that p." Or one might start with possible and define necessary as follows: "It is necessary that p if and only if it is not the case that it is possible that it is not the case that p." Both contingent and impossible, moreover, can easily be defined in terms of either possible or necessary.

Beginning in the second half of the 20th century, it became increasingly common for philosophers concerned with modality to approach the topic via the Leibnizian concept of possible worlds. Thus, a proposition is possibly true if it is true in some (at least one) possible world, necessarily true if it is true in all possible worlds, impossible if it is true in no possible world, and contingently true if it is true in the actual world but false in some possible world. Philosophers who employ the concept of possible worlds generally think of them as abstract objects, about which there are objective, mind-independent truths.

For example, assuming that time travel has been precisely defined, either there exist possible worlds in which time travel occurs or there exist no such worlds. If such worlds exist, time travel is "metaphysically" possible; if no such worlds exist, time travel is metaphysically impossible. If time travel is metaphysically possible but occurs only in worlds in which the laws of nature are different from those of the actual world, then time travel, despite its metaphysical possibility, is physically impossible. If time travel is physically possible but occurs only in worlds in which technology is far in advance of the technology of the present actual world, then time travel, despite its physical possibility, is technologically or practically impossible. Physical and technological impossibility are examples of what are called "restricted" or "relative" modalities—those that are relative to one or more specifiable factors (e.g., the laws of nature or a certain level of technology). The term "metaphysical modality" is a common way of referring to absolutely unrestricted modality, or modality that is not relativized to any particular factor.

The modality that is discussed above is the modality of propositions, or the modality of truth and falsity. The modality of propositions is sometimes thought of as being only a species of the genus modality, the other species being the modality of things. The two species are customarily designated by the medieval Latin terms de dicto (from dictum, a thing that has been said) and de re (from res, thing). The modality discussed above is therefore modality de dicto.

Whereas modality de dicto has to do with the relation of propositions to their truth-values (i.e., to truth and falsity), modality de re has to do with the relation of things to their properties (i.e., features, qualities, attributes, or characteristics). The basic concepts of modality de re are the accidental and the essential (or necessary) possession of properties: a thing has a property accidentally if it has that property in the actual world but lacks it in some possible world in which it exists, and a thing has a property essentially or necessarily if it has it in every possible world in which it exists. For example, Michelangelo's David is now in Florence, but it has that feature only accidentally, since it could at some point have been moved to some other city without ceasing to exist: there are possible worlds in which the David is in Rome or Beijing. But it seems evident that the David could not have existed without being a material thing—there is no possible world in which the David is immaterial—and that it therefore it has the property being a material thing essentially, or necessarily.

In the 20 years following World War II, owing principally to arguments of the American philosopher W.V.O. Quine, a "deflationary" theory of modality was ascendant. According to that theory, necessity is nothing more than analyticity, or the quality of a proposition whereby it is true by virtue of the meanings of the terms that are used to express it.

The deflationary theory was held to entail that necessity de re is an incoherent concept. The argument for that conclusion was supposed to have been established by Quine's famous "mathematical cyclist" argument, an adaptation of which is the following:

Suppose that the proposition "All cyclists are bipedal" is analytic. The proposition "All mathematicians are bipedal" is obviously not analytic. Given that necessity is a matter of analyticity, it follows that necessity de re, if it is a coherent notion, is a matter of analyticity. Therefore, all cyclists are necessarily bipedal, and all mathematicians are not necessarily bipedal (i.e., no mathematicians are necessarily bipedal). It follows that the American cyclist and mathematician Julia Robinson (1919–85) was necessarily bipedal (because she was a cyclist) and not necessarily bipedal (because she was a mathematician)—a contradiction. Because the notion of necessity de re thus implies a contradiction (given the deflationary theory), it is incoherent.

Since about 1970, however, most philosophers in the analytical tradition have been convinced by the work of the American philosopher Saul Kripke and others that the mathematical-cyclist argument fails. Those philosophers point out that there obviously exist possible worlds in which Robinson is not bipedal or at least ceased to be bipedal at some very early age. She might, for example, have lost a leg in an accident when she was an infant. She is, therefore, accidentally bipedal. It may be plausibly stipulated that she is bipedal in every possible world in which she is a cyclist. But she is a cyclist only accidentally, since there are possible worlds in which she is not a cyclist. Accordingly, the proposition "All cyclists are bipedal," even if it is assumed to be necessary, does not imply that all cyclists are necessarily bipedal, understood as an attribution of necessity de re. It seems, in fact, that no cyclist is necessarily bipedal: all cyclists (all people who are actually cyclists) exist in possible worlds in which they do not have two legs.

3

PLATONISM

3.1 INTRODUCTION

Platonism is the philosophy of Plato and philosophical systems closely derived from it, though contemporary platonists do not necessarily accept all of the doctrines of Plato. Platonism had a profound effect on Western thought. Platonism at least affirms the existence of abstract objects, which are asserted to exist in a third realm distinct from both the sensible external world and from the internal world of consciousness, and is the opposite of nominalism. This can apply to properties, types, propositions, meanings, numbers, sets, truth values, and so on. Philosophers who affirm the existence of abstract objects are sometimes called platonists; those who deny their existence are sometimes called nominalists. The terms "platonism" and "nominalism" also have established senses in the history of philosophy. They denote positions that have little to do with the modern notion of an abstract object.

In a narrower sense, the term might indicate the doctrine of Platonic realism, a form of mysticism. The central concept of Platonism, a distinction essential to the Theory of Forms, is the distinction between the reality which is perceptible but unintelligible, associated with the flux of Heraclitus and studied by the likes of science, and the reality which is imperceptible but intelligible, associated with the unchanging being of Parmenides and studied by the likes of mathematics. Geometry was the main motivation of Plato, and this also shows the influence of Pythagoras. The Forms are typically described in dialogues such as the Phaedo, Symposium and Republic as perfect archetypes of which objects in the everyday world are imperfect copies. Aristotle's Third Man Argument is its most famous criticism in antiquity.

In the Republic the highest form is identified as the Form of the Good, the source of all other Forms, which could be known by reason. In the Sophist, a later work, the Forms being, sameness and difference are listed among the primordial "Great Kinds". Plato established the Academy, and in the 3rd century BC, Arcesilaus adopted academic skepticism, which became a central tenet of the school until 90 BC when Antiochus added Stoic elements, rejected skepticism, and began a period known as Middle Platonism.

In the 3rd century AD, Plotinus added additional mystical elements, establishing Neoplatonism, in which the summit of existence was the One or the Good, the source of all things; in virtue and meditation the soul had the power to elevate itself to attain union with the One. Many Platonic notions were adopted by the Christian church which understood Plato's Forms as God's thoughts, while Neoplatonism became a major influence on Christian mysticism in the West through Saint Augustine, Doctor of the Catholic Church, who was heavily influenced by Plotinus' Enneads, and in turn were foundations for the whole of Western Christian thought. Many ideas of Plato were incorporated by the Roman Catholic Church.

The essence of Platonism lies in a distinction between two worlds—the familiar world of everyday life, which is the object of the senses, and an unseen world of true realities, which can be the object of the intellect. Ordinary people recognize the existence of the former and ignore that of the latter; they fail to appreciate the extent to which their beliefs both about fact and about values are arbitrarily assumed and involve internal contradictions. The philosopher is in a position to show them how insubstantial is the foundation on which they take their stand. The philosopher can demonstrate how little thought there is in popular conceptions of good and evil and can show that the very concept of sense knowledge involves difficulties, because knowledge presupposes a stable object and the objects of sense are constantly changing.

The claim, however, is that philosophers can do more than this. Because of the presence in them of something like a divine spark, they can, after suitable preparation, fix their intellectual gaze on the realities of the unseen world and, in the light of them, know both what is true and how to behave. They will not attain this result easily—to get to it will involve not only immense intellectual effort, including the repeated challenging of assumptions, but also turning their backs on everything in life that is merely sensual or animal. Yet, despite this, the end is attainable in principle, and those who arrive at it will exercise the most important part of themselves in the best way that is open to them.

That this type of view has an immediate appeal to persons of a certain kind goes without saying. There is ample evidence in poetry and elsewhere of the frequently experienced sense of the unreality of familiar things and the presence behind them of another order altogether. Platonism may be said to build on "intuitions" of this kind; as a metaphysics, its job is to give them intellectual expression, to transfer them from the level of sentiment to that of theory. It is important, however, to notice that Platonism is not just the intellectualizing of a mood. It is an attempt to solve specific problems in a specific way. In Plato's own case, the problems were set by loss of confidence in traditional morality and the emergence of the doctrine that "man is the measure of all things."

Plato thought he could counter this doctrine by appeal to another contemporary fact, the rise of science as shown in the development of mathematical knowledge. Mathematics, as he saw it, offered certain truth, although not about the familiar world. The triangle whose properties were investigated by the geometrician was not any particular triangle but the prototype that all particular triangles presuppose. The Triangle and the Circle belonged not to the world of the senses but to the world of the intelligence; they were forms. If this could be said of the objects of mathematical discourse, the same should also be true of the objects of morality. True Justice and true Goodness were not to be found in popular opinions or human institutions but should be seen as unchanging forms, eternally existing in a world apart.

Modern philosophers have found much to criticize in this system: some have objected that forms are not so much existents as abstractions, while others have found the argument from science to morality quite inconclusive because of what they allege to be an absolute dichotomy between fact and value. It may be that nobody today can subscribe to Platonism in precisely the form given it by Plato himself. The general idea, however, has certainly not lost its hold, nor have the moral perplexities to which Plato hoped to find an answer been dissipated by further thought.

3.2 PHILOSOPHY

3.2.1 Plato Holding his Timaeus, Detail from the Vatican Fresco the School of Athens

The primary concept is the Theory of Forms. The only true being is founded upon the forms, the eternal, unchangeable, perfect types, of which particular objects of moral and responsible sense are imperfect copies. The multitude of objects of sense, being involved in perpetual change, are thereby deprived of all genuine existence. The number of the forms is defined by the number of universal concepts which can be derived from the particular objects of sense. The following excerpt may be representative of Plato's middle period metaphysics and epistemology:

"Since the beautiful is opposite of the ugly, they are two."

"Of course."

"And since they are two, each is one?"

"I grant that also."

"And the same account is true of the just and unjust, the good and the bad, and all the forms. Each of them is itself one, but because they manifest themselves everywhere in association with actions, bodies, and one another, each of them appears to be many."

"That's right."

"So, I draw this distinction: On one side are those you just now called lovers of sights, lovers of crafts, and practical people; on the other side are those we are now arguing about and whom one would alone call philosophers."

"How do you mean?"

"The lovers of sights and sounds like beautiful sounds, colors, shapes, and everything fashioned out of them, but their thought is unable to see and embrace the nature of the beautiful itself."

"That's for sure."

"In fact, there are very few people who would be able to reach the beautiful itself and see it by itself. Isn't that so?"

"Certainly."

"What about someone who believes in beautiful things, but doesn't believe in the beautiful itself and isn't able to follow anyone who could lead him to the knowledge of it? Don't you think he is living in a dream rather than a wakened state? Isn't this dreaming: whether asleep or awake, to think that a likeness is not a likeness but rather the thing itself that it is like?"

"I certainly think that someone who does that is dreaming."

"But someone who, to take the opposite case, believes in the beautiful itself, can see both it and the things that participate in it and doesn't believe that the participants are it or that it itself is the participants--is he living in a dream or is he awake?

"He's very much awake."

Book VI of the Republic identifies the highest form as the Form of the Good, the cause of all other Ideas, and that on which the being and knowing of all other Forms is contingent. Conceptions derived from the impressions of sense can never give us the knowledge of true being; i.e. of the forms. It can only be obtained by the soul's activity within itself, apart from the troubles and disturbances of sense; that is to say, by the exercise of reason. Dialectic, as the instrument in this process, leading us to knowledge of the forms, and finally to the highest form of the Good, is the first of sciences. Later Neoplatonism, beginning with Plotinus, identified the Good of the Republic with the transcendent, absolute One of the first hypothesis of the Parmenides (137c-142a).

Platonist ethics is based on the Form of the Good. Virtue is knowledge, the recognition of the supreme form of the good. And, since in this cognition, the three parts of the soul, which are reason, spirit, and appetite, all have their share, we get the three virtues, Wisdom, Courage, and Moderation. The bond which unites the other virtues is the virtue of Justice, by which each part of the soul is confined to the performance of its proper function.

Platonism had a profound effect on Western thought. In many interpretations of the Timaeus Platonism, like Aristotelianism, poses an eternal universe, as opposed to the nearby Judaic tradition that the universe had been created in historical time, with its continuous history recorded. Unlike Aristotelianism, Platonism describes idea as prior to matter and identifies the person with the soul. Many Platonic notions secured a permanent place in Christianity.

At the heart of Plato's philosophy is the theory of the soul. Francis Cornford described the twin pillars of Platonism as being the theory of the Forms, on the one hand, and, on the other hand, the doctrine of the immortality of the soul. Indeed, Plato was the first person in the history of philosophy to believe that the soul was both the source of life and the mind. In Plato's dialogues, we find the soul playing many disparate roles. Among other things, Plato believes that the soul is what gives life to the body in terms of self-motion: to be alive is to be capable of moving yourself; the soul is a self-mover. He also thinks that the soul is the bearer of moral properties. The soul is also the mind: it is that which thinks in us.

We see this casual oscillation between different roles of the soul in many dialogues. First of all, in the Republic:

Is there any function of the soul that you could not accomplish with anything else, such as taking care of something (epimeleisthai), ruling, and deliberating, and other such things? Could we correctly assign these things to anything besides the soul, and say that they are characteristic (idia) of it?

No, to nothing else.

What about living? Will we deny that this is a function of the soul?

That absolutely is.

The Phaedo most famously caused problems to scholars who were trying to make sense of this aspect of Plato's theory of the soul, such as Broadie and Dorothea Frede.

More-recent scholarship has overturned this accusation arguing that part of the novelty of Plato's theory of the soul is that it was the first to unite the different features and powers of the soul that became commonplace in later ancient and medieval philosophy. For Plato, the soul moves things by means of its thoughts, as one scholar puts it, and accordingly, the soul is both a mover (i.e., the principle of life, where life is conceived of as self-motion) and a thinker.

3.3 HISTORY OF PLATONISM

3.3.1 Ancient Philosophy

The Academy

Platonism was originally expressed in the dialogues of Plato, in which the figure of Socrates is used to expound certain doctrines, that may or may not be similar to the thought of the historical Socrates, Plato's master. Plato delivered his lectures at the Platonic Academy, a precinct containing a sacred grove outside the walls of Athens. The school continued there long after Plato's death. There were three periods: the Old, Middle, and

New Academy. The chief figures in the Old Academy were Speusippus (Plato's nephew), who succeeded him as the head of the school (until 339 BC), and Xenocrates (until 313 BC). Both of them sought to fuse Pythagorean speculations on number with Plato's theory of forms.

3.4 THE SKEPTICAL ACADEMY

Around 266 BC, Arcesilaus became head of the Academy. This phase, known as the Middle Academy, strongly emphasized philosophical skepticism. It was characterized by its attacks on the Stoics and their assertion of the certainty of truth and our knowledge of it. The New Academy began with Carneades in 155 BC, the fourth head in succession from Arcesilaus. It was still largely skeptical, denying the possibility of knowing an absolute truth; both Arcesilaus and Carneades argued that they were maintaining a genuine tenet of Plato.

3.4.1 Middle Platonism

Around 90 BC, Antiochus of Ascalon rejected skepticism, making way for the period known as Middle Platonism, in which Platonism was fused with certain Peripatetic and many Stoic dogmas. In Middle Platonism, the Platonic Forms were not transcendent but immanent to rational minds, and the physical world was a living, ensouled being, the World-Soul. Pre-eminence in this period belongs to Plutarch. The eclectic nature of Platonism during this time is shown by its incorporation into Pythagoreanism (Numenius of Apamea) and into Jewish philosophy.

3.4.2 Neoplatonism

Many Western churchmen, including Augustine of Hippo, have been influenced by Platonism. In the third century, Plotinus recast Plato's system, establishing Neoplatonism, in which Middle Platonism was fused with mysticism. At the summit of existence stands the One or the Good, as the source of all things. It generates from itself, as if from the reflection of its own being, reason, the nous, wherein is contained the infinite store of ideas. The world-soul, the copy of the nous, is generated by and contained in it, as the nous is in the One, and, by informing matter in itself nonexistent, constitutes bodies whose existence is contained in the world-soul. Nature therefore is a whole, endowed with life and soul. Soul, being chained to matter, longs to escape from the bondage of the body and return to its original source. In virtue and philosophical thought it has the power to elevate itself above the reason into a state of ecstasy, where it can behold, or ascend to, that one good primary Being whom reason cannot know. To attain this union with the Good, or God, is the true function of human beings.

Plotinus' disciple, Porphyry, followed by Iamblichus, developed the system in conscious opposition to Christianity. The Platonic Academy was re-established during this period; its most renowned head was Proclus (died 485), a celebrated commentator on Plato's writings.

3.5 MEDIEVAL PHILOSOPHY

3.5.1 Christianity and Platonism

Platonism has had some influence on Christianity through Clement of Alexandria and Origen, and the Cappadocian Fathers. St. Augustine was heavily influenced by Platonism as well, which he encountered through the Latin translations of Marius Victorinus of the works of Porphyry and/or Plotinus.

Platonism was considered authoritative in the Middle Ages. Platonism also influenced both Eastern and Western mysticism. Meanwhile, Platonism influenced various philosophers. While Aristotle became more influential than Plato in the 13th century, St. Thomas Aquinas's philosophy was still in certain respects fundamentally Platonic.

3.5.2 Modern Philosophy

Renaissance

The Renaissance also saw a renewed interest in Platonic thought, including more interest in Plato himself. In 16th-, 17th-, and 19th-century England, Plato's ideas influenced many religious thinkers including the Cambridge Platonists. Orthodox Protestantism in continental Europe, however, distrusts natural reason and has often been critical of Platonism. An issue in the reception of Plato in early modern Europe was how to deal with the same-sex elements of his corpus.

Christoplatonism is a term used to refer to a dualism opined by Plato, which holds spirit is good but matter is evil, which influenced some christian churches, though the Bible's teaching directly contradicts this philosophy and thus it receives constant criticism from many teachers in the Christian Church today. According to the Methodist Church, Christoplatonism directly "contradicts the Biblical record of God calling everything He created good."

3.6 CONTEMPORARY PHILOSOPHY

3.6.1 Modern Platonism

Apart from historical Platonism originating from thinkers such as Plato and Plotinus, we also encounter the theory of abstract objects in the modern sense.

Platonism is the view that there exist such things as abstract objects — where an abstract object is an object that does not exist in space or time and which is therefore entirely non-physical and non-mental. Platonism in this sense is a contemporary view.

Most contemporary Platonists trace their views to those of Gottlob Frege.

This modern Platonism has been endorsed in one way or another at one time or another by numerous philosophers, such as Bernard Bolzano, who argue for anti-psychologism. Plato's works have been decisively influential for 20th Century Philosophers such as

Alfred North Whitehead and his Process Philosophy; and for the critical realism and metaphysics of Nicolai Hartmann.

3.6.2 Analytic

In contemporary philosophy, most Platonists trace their ideas to Gottlob Frege's influential paper "Thought", which argues for Platonism with respect to propositions, and his influential book, The Foundations of Arithmetic, which argues for Platonism with respect to numbers and is a seminal text of the logicist project. Contemporary analytic philosophers who espoused Platonism in metaphysics include Bertrand Russell, Alonzo Church, Kurt Gödel, W. V. O. Quine, David Kaplan, Saul Kripke, Edward Zalta and Peter van Inwagen. Iris Murdoch espoused Platonism in moral philosophy in her 1970 book The Sovereignty of Good. Paul Benacerraf's epistemological challenge to contemporary Platonism has proved its most influential criticism.

Continental

In contemporary Continental philosophy, Edmund Husserl's arguments against psychologism are believed to derive from a Platonist conception of logic, influenced by Frege and his mentor Bolzano. —Husserl explicitly mentioned Bolzano, G. W. Leibniz and Hermann Lotze as inspirations for his position in his Logical Investigations (1900–1). Other prominent contemporary Continental philosophers interested in Platonism in a general sense include Leo Strauss, Simone Weil, and Alain Badiou.

3.7 THE BACKGROUND TO PLATO'S METAPHYSICS

Three predecessors heavily influenced Plato's thoughts on metaphysics and epistemology, Heraclitus (c. 540 B.C.–480–70), Parmenides (c.515 B.C.–449–40), and Socrates (470 B.C.–399). Only fragments remain of the writings of Parmenides and Heraclitus, including some contained in the dialogues of Plato. Socrates wrote nothing. Plato's depiction of his teacher is our primary source of evidence for his philosophy. Parmenides argued that there is and could be only one thing, Being. One could not even think or say what is not. Moreover, since change implies that something comes to be what it was not—I change from not being tan to being tan, nothing can change. Reality is static. The appearance of change is just that, a deceptive appearance. Unfortunately, what little we have left of Parmenides does not allow us to decide whether he argued that there is just one item, Being, in his universe—strict numerical monism—or whether there is just one kind of thing, beings or things that are. Parmenides' account of Being seems to have contributed to Plato's doctrine of Forms.

Heraclitus is the apostle of change. For Heraclitus, the ordinary objects of the physical world seem to be continually changing. The only constant, the underlying commonality, is the pattern of change itself. That there are entities that do not change is, for Heraclitus, an illusion. Heraclitus' notion of 'flux' seems to have influenced Plato's thinking about ordinary material objects.

In the opinion of most scholars, the seminal influence on all of Plato's thinking was Socrates. However, it appears from the writings of Plato, as well as those of the historian Xenophon and the comic poet Aristophanes, that Socrates was almost exclusively interested in ethics. This is not to say that metaphysical or epistemological issues were of no concern to him. Rather, these sources convey the impression that Socrates was not particularly interested in articulating a metaphysical or epistemological theory. Rather, concerned with caring for the soul so that one might live happily, he uses both epistemological and metaphysical theses in search of answers to his ethical questions. However, it is not easy to distinguish when one is engaged in metaphysical theorizing from when is merely using metaphysical notions. The claim that Socrates was not a metaphysician or epistemologist is particularly hard to evaluate, for we have basically only Plato's dialogues as evidence. Since Plato uses Socrates as a mouthpiece in many of his writings, readers are forced to ask when or whether one is reading the doctrines of Socrates, or Plato, or neither. This 'Socratic question' is intimately involved with the question of Plato's development and the chronology of his dialogues.

In all likelihood, Plato wrote different dialogues at different times. We typically divide his writings into three periods. In the early 'Socratic' period, we find Apology, Crito, Euthyphro, Charmides, Ion, Lysis, Laches, Hippias Minor, Menexenus, Euthydemus and the Protagoras. The Hippias Major, Gorgias and perhaps the Meno belong to the end of this period, maybe with the Gorgias and more likely the Meno verging into the middle period. The middle period works include the Cratylus, Symposium, Phaedo, Republic and perhaps the Phaedrus. In the post-Republic phase we then find the Parmenides, Theaetetus, Sophist, Politicus, Timaeus, Philebus and Laws, along with the Critias. The Socratic dialogues, so-called because Socrates is always the chief interlocutor, are thought to present doctrines of Socrates himself. These are dialogues devoted to ethical inquiries into the virtues, e.g., what is courage, or what is justice? In contrast, the middle period dialogues are thought to present the views of Plato, though nonetheless Socrates remains the speaker. Here for the first time we find remarks about the immortality of the soul, about special entities called 'Forms' that exist outside of space and time and that are both the objects of knowledge and somehow the cause of whatever transpires in the physical world, and the doctrine of recollection, the thesis that the immortal soul, in a disembodied state prior to its incarceration in a body, viewed these Forms, knowledge of which is then recalled by incarcerated souls through a laborious process. Socrates, in the early Apology, is non-committal about the immortality of the soul. Similarly, in the early dialogues we find that Socrates, in keeping with the claim that he is neither a metaphysician nor epistemologist, has nothing to say about recollection and never explicitly appeals to Forms. It is thus in the middle period works that one locates Plato's first thoughts about epistemological and metaphysical issues. To those topics we shall turn shortly. But these are, in the eyes of many, just first thoughts; for the dialogues in the late period suggest changes to key ethical, epistemological and metaphysical doctrines found in these middle period works. Over the course of the last fifty years, scholars have debated whether and to what extent

Plato changed his views. The debate has grown so involved that it is perhaps best not to worry whether anyone believes the extreme positions that, on the one hand, Plato conceived of every one of his major doctrines before he ever wrote, or, on the other hand, that he changed his mind on central theses from one dialogue to the next. Broadly speaking, those who maintain that Plato keeps to his central theses from one period to the next are Unitarians. Those who believe that he changes his views from one period to the next are Developmentalists. The most plausible position, and the perhaps the dominant position in the contemporary scholarship, is somewhere in the middle. About some theses, Plato, over the course of his writings, expands his thoughts, recognizes difficulties, and even changes his mind. About other theses he stands by his fundamental insights.

A prime example of the interpretative problems facing the student of Plato is the development of his most distinctive doctrine, the theory of Forms. Aristotle, in recounting Plato's intellectual development, reports that "Socrates was the first to seek the universal in ethical matters but that he did not separate it. Plato, marrying Socrates' philosophy with that of Heraclitus, separated the universal, on the grounds that the sensible order, where Socrates had focused, was in flux." Plato, Aristotle tells us, called these separated universals 'Forms'. Universal is a technical notion in metaphysics: a universal is that which is predicable of many. It is meant to capture the intuition that a variety of things can all have the same feature or property. For instance, a bowling ball, a basketball, and a figure drawn on a blackboard can all be round. What many things have in common, or a feature they share, is a universal or, in Plato's terms, a Form. Of course there seems to be a huge number of properties. Many different things are white. Many different things are animals. Each (shared) property is a universal—a 'one over many instances,' whiteness over the many white things, roundness over the many round things, and so on. Thus, for Plato, Roundness and Whiteness are Forms. Following the lead of Aristotle, scholars have focused on what it means for Plato, in contrast to Socrates, to have separated his universals, the Forms. The starting point, then, for the study of Plato's metaphysics, is the Socratic dialogues and Socrates' investigation into universals of the ethical variety, namely Justice, Piety, Courage and others.

In the early dialogues, Socrates seeks these ethical universals through a distinctive mode of inquiry, the 'elenchus'. Elenctic inquiry is fundamentally a form of cross-examination, where Socrates tries to elicit from others their beliefs about matters of justice or piety, etc. Typically the result is that his interlocutors turn out to have an inconsistent set of beliefs about the virtues. At the heart of the Socratic elenchus is the 'What is X' question (where 'X' typically names an ethical property). The answers offered to these questions fail usually because they are too narrow or too wide. An answer is too narrow if it fails to include all cases. An answer is too wide if, while it includes all cases of, for instance, piety, it also includes other things, cases of justice or impiety. We can infer from these failed definitions a set of conditions Socrates places on an adequate answer to his 'What is X' questions. He is seeking an answer which picks out a Socratic Property, e.g., Piety, that is a universal such that: it is found whenever and wherever there is an instance of Piety; and it 'causes' or 'makes' the instance to be such as it is. Piety's power to make, e.g., Socrates, pious derives from Piety's itself being pious. Piety self-predicates: Piety is pious. Because it is pious, when Piety is present to/in Socrates, Piety causes Socrates to be pious. In the Socratic dialogues Plato does not distinguish the (metaphysical) way in which Socrates is pious from the way in which Piety is pious—in these dialogues there appears to be just one ontological predication relation. One has knowledge of a Socratic Property when she can give an account (logos) that says what X is, that is, when she can give the definition of the property under investigation. Treating a definition as a linguistic item, we can say that the definition specifies or picks out the essence (ousia) of the property, and a definitional statement predicates the essence of the property whose essence it is. It is unclear from the Socratic dialogues whether any other property is predicated of a Socratic Property: arguably Piety is pious and only pious. In contrast, the things that are pious, e.g., Socrates or saying a prayer, have many properties. From what we can infer from Plato's remarks in these early dialogues, and from Aristotle's remarks, a Socratic property is in the sensibles—It is an immanent universal. In this respect, the essence of Piety is also found in Socrates (and thus the linguistic definition of Piety is also linguistically predicable of Socrates). Plato's distinctive 'separation of the universal' might then be viewed as his rejection of Socrates' assumption that the universal is in the sensibles, (and accordingly a rejection of Socrates' assumption that the definition is predicable, or predicable in the same way, of the sensible instances as well as the universal.) If Aristotle is right, Plato's problem with sensibles is that they change.

Some philosophers and philosophies are defined most of all by their methods. Some are defined by their results. Plato's philosophy is remarkable for both. Blake Hestir's book admirably attends to the interaction between one of Plato's most important argumentative strategies and the substantive results of deploying that strategy. Plato relies on what Hestir calls "grounding arguments" to defend the view that the possibilities of meaning, predication and truth require entities that are stable (i.e. one and the same in at least one respect) yet also complex and combined to yield a "dynamically structured", "eternally actualizing" metaphysical foundation. According to Hestir, Plato's view that truth is a substantive property of statements finds its support in his realism about being. Identifying and understanding the metaphysical foundations of truth, as Plato sees them, is the primary undertaking of Hestir's book.

As his interpretation develops, Hestir posits a revision in Plato's metaphysics from the "strong Platonism" of Phaedo, Republic and Symposium to the "restricted Platonism" of Cratylus, Theaetetus, Parmenides and Sophist. According to the ontology of strong Platonism, Forms are the fundamental beings, and Forms are: simply one (not many); always the same in the same respect; beings in every way, in no way admitting of not being; incapable of combination with other Forms. Restricted Platonism, on the other hand, posits forms that are: one in at least one respect, but not every respect; always the same in at least one respect, but not every respect; beings in at least one respect, but not every respect; beings which, as such, must combine with other forms. Plato is said to replace isolated, "pure" strong Platonic Forms with interactive beings that can combine with one another while remaining stable at least with respect to what they are, even if they are not stable in every respect. What motivates the shift away from strong Platonism? According to Hestir, repeated applications of grounding arguments reveal that strong Platonism can be replaced with the "more defensible ontology" of restricted Platonism.

Hestir locates the following paradigmatic grounding argument at Parmenides 135b-c:

GAParm1. Forms (eide) are necessary for the possibility of meaningful discourse (dialegesthai dunamin).

GAParm2. Meaningful discourse is possible.

GAParm3. So there are forms.

Support for the premises initially comes from arguments in Cratylus and Theaetetus. In those dialogues, Socrates critically assesses a radical Heraclitean position according to which everything always moves in every way, nothing rests, and nothing is one. According to Hestir, Socrates argues that, under Heraclitean conditions, reference and predication (and being and knowledge) would be impossible. Meaningful discourse, Socrates infers, requires subjects and properties that are stable (i.e. one and the same) in at least one respect, even if not in all respects. These are heavily contested passages and Hestir skillfully navigates textual difficulties along with controversies among commentators. The discussion is thoughtful and rewarding even for those who will disagree with Hestir's reconstruction of the arguments.

Like Hestir, I read Plato as self-consciously relying on grounding arguments in order to defend basic tenets of his metaphysics and semantics. I also agree with Hestir that some of the most important grounding arguments fall short of defending strong Platonism. Even so, I am less confident that the arguments Hestir considers from the Cratylus and Theaetetus are intended to support restricted Platonism directly, or are successful in defending any kind of Platonism at all. The sort of stability required for referential and predicative success might very well be met by, for example, Aristotelian particulars and universals. I am not sure that Hestir would disagree with this last point. At times, he remarks that Plato claims more for his grounding arguments than they actually establish. But at other points in the book he is surprisingly sanguine about attributing to Plato the view that it is built into the stability required for merely meaningful discourse that the requisite beings are Platonic objects of definition and knowledge. He says:

For Plato, the condition of being the same in at least one respect is being always the same f, being that f that explains what it is to be f. And as far as Plato is concerned, none of the many f-things can explain what it is to be f even if there only one f-thing and it is f for a very long time.

If Hestir is right, then the first premise of the Parmenides' grounding argument is far more controversial than it might appear at first glance. Hestir recognizes that it is a "significant argumentative leap" to require the eternal stability of Platonic forms for referential and predicative success. Yet this leap is called upon to support Hestir's fairly deflationary reading of dialegesthai dunamin as the capacity for "meaningful discourse." If Plato instead wished to ground the possibility of full-blown dialectic in the Parmenides, requiring the "definitional stability" Platonic explanatory entities could provide would make more sense. Hestir might very well be right about the various attributions to Plato and their implications for Plato's semantics, but I'm not sure that he is, and I expect that readers of Part I would benefit from a more expanded discussion of why Plato is tempted by the argumentative leap and how he might defend the leap, if, indeed, he makes it.

While a good portion of the book considers the conditions on the possibility of meaningful discourse, the bulk of it and its heart, it seems to me is occupied with articulating what it is for a form (or any being) to have being, either in virtue of itself or with respect to something else. Since truth depends on being, the book strives to characterize Plato's commitments in the Sophist to being as a "greatest kind" and to the various roles that being plays in structuring reality. I cannot do justice here to the insightful and detailed arguments in these chapters of the book. I will mention only a few of their many enticing findings.

Hestir argues that being is a fundamental, atypical kind; it just is the kind that functions to make capacities and their realization in combination actual. Beings, as such, have or are capacities to affect and to be affected. According to Hestir, "the kind being just is capacity and the realization of that capacity in combinations". Moreover, being runs through all beings. A form's internal combination with the kind being is a precondition of its further capacity to combine with any other being. Hestir claims that "what needs to be internally 'held together' is a form's sameness, difference, and essential content". Unified, well-individuated beings are then eligible to enter into relations and combinations with other individuated beings. The kind being, as capacity and combination, grounds all ontological unity, stability and combination. Understanding being in this way helps to explain, in part, Plato's claim that "it is the nature of the kinds to commune with each other".

With a conception of the metaphysical structure of forms in place, we are well-situated to appreciate the first premise of a second important grounding argument from Sophist 259d-e:

GASoph1. The combination of forms is necessary for the possibility of statement.

GASoph2. Statement is possible.

GASoph3. So the forms combine.

Hestir proposes that the investigation of the kinds being, difference and same in the Sophist reveals, first, that forms must combine with other forms if they are to exist at all, and, second, that forms must combine with other beings in ontic predications to ground true statements.

An efficient formulation, and one which displays the various connections of interest to Hestir, can be found in the following account of truth for a statement:

The statement 's is f' is true ↔ the form signified by 'f' has being in the case of the subject named by 's'.

Truth requires, first of all, a statement, something truth-evaluable. A simple list of words is not a candidate for being true. The Sophist proposes that some types of terms, names, must be capable of fitting together with other types of terms, verbs or predicates, to produce statements. Statements "accomplish something" by asserting or denying. Both 'Theaetetus is sitting' and 'Theatetus is flying' are statements. Both are realizations of the capacities of certain words to combine. But one is true and the other false. The first has the property of being true, since in its case the entities designated by the words in the statement are in fact combined as the statement asserts. The second is false since, although it successfully asserts something, the metaphysical conditions on truth are not satisfied.

'Theaetetus is sitting' is true if and only if the form sitting has being in the case of Theaetetus. Hestir explains: that sitting is available as a form means that it must be combined with at least being, same, and different prior to entering into a relation with Theaetetus. Such combination is necessary for sitting being a being, being self-identical and being different from other things. That sitting holds of Theatetus requires these combinations in addition to the straightforward participation of Theaetetus in sitting.

In the example, truth depends on 'sitting' signifying the form sitting, 'Theaetetus' naming Theatetus and on the form sitting having being relative to the perceptible Theaetetus. The account of truth which Hestir reconstructs accords priority to properties of subjects, to the forms predicated of a subject. Since the form f makes something else to be f, the f's having being relative to s is the truthmaking condition for 's is f.'

For Plato, truth depends on being. Statements are true in virtue of the world being a certain way, in virtue of highly complex metaphysical structures and relations. The kind being ensures that forms exist and makes possible the combination of forms with other beings in ontic predications. But, according to Hestir, the Sophist's substantive realist conditions on truth do not require additional, "seriously dyadic" relations of correspondence or congruence between statements and facts. Truth requires structured being. But, Hestir suggests, truth is not defined in terms of any isomorphism or mirroring relation between statements and reality.

Blake Hestir's scholarship has consistently encouraged readers of Plato and Aristotle to pay closer attention to the rich, sometimes unexpected details of their conceptions of truth and falsity. This book is an accomplished, welcome extension of Hestir's efforts to date. I especially appreciated the impressive engagement with the Sophist's metaphysics and the patient accounting of Plato's conception of truth. For students of Plato's semantics and metaphysics more generally, Hestir's book offers lots of food for thought. I highly recommend it.

3.8 THE METAPHYSICS OF THE PHAEDO

The Phaedo is Plato's eulogy to Socrates. It recounts the last hours of Plato's teacher. Socrates/Plato wants to convince us that we should care about our souls and that the best way to care for the soul is to live philosophically. Towards that end we find a series of arguments whose aim is to prove the immortality of the soul. At least three of these arguments, the Argument from Recollection and its prelude (65a-67a and 72e-78b), the Affinity Argument (78b-84b), and the Final Argument (102a-107a) and its prelude (95a-102a), are crucial for understanding Plato's initial thoughts on metaphysics and epistemology. Here Plato draws a contrast between unchanging Forms and changing material particulars. Unfortunately, neither in the Phaedo nor in any other dialogue do we find Plato giving a detailed description of the nature of Forms, or particulars, or their interaction. What is referred to as Plato's theory of Forms is thus a rational reconstruction of Plato's doctrine. In such a reconstruction scholars try to determine a set of principles or theses which, taken together, allow us to show why Plato says what he does about Forms, souls, and other metaphysical items. In the attempt to make more precise what Plato is after, one risks attributing to Plato notions that are either not his or not as well developed in Plato as scholars would hope. Perhaps the notion of a particular is such a case. Intuitively, particulars are things like my dog Ajax, Venus, my computer, and so on, the ordinary material things of the everyday spatio-temporal world. In a rational reconstruction, we can be more precise by stipulating, for instance, that a particular is that of which properties are predicated and which is never predicated of anything. This 'stipulated definition' is nowhere in Plato, though it may well capture his thinking about ordinary particulars. For the sake of exposition, I will assume that in the Phaedo Plato is appealing to our naïve, intuitive understanding of what it is for something to be a material particular. In the author's opinion, the metaphysics of the Phaedo and other middle period works is devoted to developing the account of Forms; perhaps because while most of us think that included in what there is are the various, e.g., dogs, people, mountains and trees, few of us ever think about whether there is some universal/Form, Justice Itself, The Large Itself, and so on, that exists outside of space and time.

The argument of the Phaedo begins from Plato's assertion that the soul seeks freedom from the body so that it may best grasp truth, because the body hinders and distracts it: the soul comes to be separate (choris) from the body, itself by itself. The senses furnish no truth; those senses about the body are neither accurate nor clear. The soul reckons best when it is itself by itself, i.e., not in contact with body (65a-65d3). At this juncture, Socrates changes course:

What about these things? Do we say that justice itself is something? Of course. And the fair and the good? Surely. Then have you ever seen any of these sorts of things with your eyes? In no way. But then have you grasped them with any other sense through the body. I am talking about all (of them), for instance about size, health, strength, in a word about the essence (ousia) of all of them, what each happens to be. Is it through the body then that what is most true of these things is contemplated? Or does it hold thus? Whoever of us should prepare himself to consider most accurately each thing itself about which he inquires, that one would come closest to knowing each thing. … he would do this most cleanly …who using his intellect itself by itself, unmixed would undertake to hunt down each of the beings, itself by itself unmixed (65d4–66a3).

This is the first passage in the dialogues widely agreed to introduce Forms. First, Forms are marked as auto kath auto beings, beings that are what they are in virtue of themselves. In subsequent arguments we learn other features of these Forms. From the next argument concerning Recollection, Forms are said to be perfect and what particulars strive to be like but fall short of. Then in the Affinity Argument we discover that Forms are simple or incomposite, of one form (monoeidetic), whereas particulars are complex, divisible and of many forms. In the crucial Final Argument, Plato finally presents the hypothesis of Forms to explain coming into being and destruction, in general, i.e., change. Once Cebes accepts the hypothesis, a novel implication is announced:

Well then, consider what then follows if you also accept my hypothesis. For it seems to me that if anything else is beautiful besides Beauty Itself, it is beautiful on account of nothing else than because it partakes of Beauty Itself. And I speak in the same way about everything else. Do you accept this sort of cause or explanation?

In this passage, Plato introduces two predication relations, Being and Partaking. Understanding how Being and Partaking function to 'tie' together the various subjects and properties mentioned in his metaphysical discussions is crucial to reconstructing his metaphysics and epistemology. At first blush, it seems that there are two kinds of subjects of which properties are predicated, namely Forms and material particulars. Similarly, at first blush it seems that there are Forms for every property involved in the changes afflicting material particulars. For instance, since particulars, e.g. Helen of Troy, change from being not-beautiful to being beautiful, there is the Form Beauty Itself. In this passage, Plato asserts that particulars like Helen, because she is not the Form but rather is a material particular, is related to, or 'tied to' Beauty in virtue of what he calls 'partaking.' Beauty Itself, on the other hand, not being something 'other than Beauty', does not partake of Beauty—it simply is beautiful. Generalizing from what is said here about Beauty Itself, it seems that Forms inherit from the Socratic Properties their self-predicational status: Beauty is beautiful; Justice is just; Equality is equal. Partaking in Beauty makes Helen beautiful because Beauty Itself is beautiful. Call this way in which a Form is related to the property it is 'Being.' Understanding Being, the way in which Beauty is beautiful, that is, determining what it is for a Form to self-predicate, is central to understanding Plato's Theory of Forms and his middle period metaphysics.

3.8.1 The Philosophical Significance of Mathematical Platonism

Mathematical platonism has considerable philosophical significance. If the view is true, it will put great pressure on the physicalist idea that reality is exhausted by the physical. For platonism entails that reality extends far beyond the physical world and includes objects which aren't part of the causal and spatiotemporal order studied by the physical sciences. Mathematical platonism, if true, will also put great pressure on many naturalistic theories of knowledge. For there is little doubt that we possess mathematical knowledge. The truth of mathematical platonism would therefore establish that we have knowledge of abstract objects. This would be an important discovery, which many naturalistic theories of knowledge would struggle to accommodate.

Although these philosophical consequences are not unique to mathematical platonism, this particular form of platonism is unusually well suited to support such consequences. For mathematics is a remarkably successful discipline, both in its own right and as a tool for other sciences. Few contemporary analytic philosophers are willing to contradict any of the core claims of a discipline whose scientific credentials are as strong as those of mathematics. So if philosophical analysis revealed mathematics to have some strange and surprising consequences, it would be unattractive simply to reject mathematics. A form of platonism based on a discipline whose scientific credentials are less impressive than those of mathematics would not be in this fortunate situation. For instance, when theology turns out to have some strange and surprising philosophical consequences, many philosophers do not hesitate to reject the relevant parts of theology.

Object Realism

Let object realism be the view that there exist abstract mathematical objects. Object realism is thus just the conjunction of Existence and Abstractness. Object realism stands opposed to nominalism, which in contemporary philosophy is typically defined as the view that there are no abstract objects.

Because object realism leaves out Independence, this view is logically weaker than mathematical platonism. The philosophical consequences of object realism are thus not as strong as those of platonism. Many physicalists would accept non-physical objects provided that these are dependent on or reducible to physical objects. They may for instance accept objects such as corporations, laws, and poems, provided that these are suitably dependent or reducible to physical objects. Moreover, there appears to be no mystery about epistemic access to non-physical objects that we have somehow made or 'constituted'. If corporations, laws, and poems are made or 'constituted' by us, presumably we gain knowledge of them in the process of making or 'constituting' them.

Some views in the philosophy of mathematics are object realist without being platonist. One example are traditional intuitionist views, which affirm the existence of mathematical objects but maintain that these objects depend on or are constituted by mathematicians and their activities.

Truth-Value Realism

Truth-value realism is the view that every well-formed mathematical statement has a unique and objective truth-value that is independent of whether it can be known by us and whether it follows logically from our current mathematical theories. The view also holds that most mathematical statements that are deemed to be true are in fact true. So truth-value realism is clearly a metaphysical view. But unlike platonism it is not an ontological view. For although truth-value realism claims that mathematical statements have unique and objective truth-values, it is not committed to the distinctively platonist idea that these truth-values are to be explained in terms of an ontology of mathematical objects.

Mathematical platonism clearly motivates truth-value realism by providing an account of how mathematical statements get their truth-values. But the former view does not entail the latter unless further premises are added. For even if there are mathematical objects, referential and quantificational indeterminacy may deprive mathematical statements of a unique and objective truth-value. Conversely, truth-value realism does not by itself entail Existence and thus implies neither object realism nor platonism. For there are various accounts of how mathematical statements can come to possess unique and objective truth-values which do not posit a realm of mathematical objects.

In fact, many nominalists endorse truth-value realism, at least about more basic branches of mathematics, such as arithmetic. Nominalists of this type are committed to the slightly odd-sounding view that, although the ordinary mathematical statement.

(1) There are primes numbers between 10 and 20.

is true, there are in fact no mathematical objects and thus in particular no numbers. But there is no contradiction here. We must distinguish between the language LM in which mathematicians make their claims and the language LP in which nominalists and other philosophers make theirs. The statement is made in LM. But the nominalist's assertion that is true but that there are no abstract objects is made in LP. The nominalist's assertion is thus perfectly coherent provided that is translated non-homophonically from LM into LP. And indeed, when the nominalist claims that the truth-values of sentences of LM are fixed in a way which doesn't appeal to mathematical objects, it is precisely this sort of non-homophonic translation she has in mind. The view mentioned in the previous note provides an example.

This shows that for the claim Existence to have its intended effect, it must be expressed in the language LP used by us philosophers. If the claim was expressed in the language

LM used by mathematicians, then nominalists could accept the claim while still denying that there are mathematical objects, contrary to the purpose of the claim.

A small but important tradition of philosophers urge that the debate about platonism should be replaced by, or at least transformed into, a debate about truth-value realism. One reason offered in support of this view is that the former debate is hopelessly unclear, while the latter is more tractable. Another reason offered is that the debate about truth-value realism is of greater importance to both philosophy and mathematics than the one about platonism.

3.8.2 The Mathematical Significance of Platonism

Working realism is the methodological view that mathematics should be practiced as if platonism was true. This requires some explanation. In debates about the foundations of mathematics platonism has often been used to defend certain mathematical methods, such as the following:

i. Classical first-order (or stronger) languages whose singular terms and quantifiers appear to be referring to and ranging over mathematical objects. (This contrasts with the languages that dominated earlier in the history of mathematics, which relied more heavily on constructive and modal vocabulary.)
ii. Classical rather than intuitionistic logic.
iii. Non-constructive methods (such as non-constructive existence proofs) and non-constructive axioms (such as the Axiom of Choice).
iv. Impredicative definitions (that is, definitions that quantify over a totality to which the object being defined would belong).
v. 'Hilbertian optimism', that is, the belief that every mathematical problem is in principle solvable.

According to working realism, these and other classical methods are acceptable and available in all mathematical reasoning. But working realism does not take a stand on whether these methods require any philosophical defense, and if so, whether this defense must be based on platonism. In short, where platonism is an explicitly philosophical view, working realism is first and foremost a view within mathematics itself about the correct methodology of this discipline. Platonism and working realism are therefore distinct views.

However, there may of course be logical relations between the two views. Given the origin of working realism, it is not surprising that the view receives strong support from mathematical platonism. Assume that mathematical platonism is true. Then clearly the language of mathematics ought to be as described in (i) Secondly, provided it is legitimate to reason classically about any independently existing part of reality, (ii) would also follow. Thirdly, since platonism ensures that mathematics is discovered rather than invented, there would be no need for mathematicians to restrict themselves to constructive methods and axioms, which establishes (iii). Fourth, there is a powerful and influential argument due to Gödel (1944) that impredicative definitions are legitimate whenever the objects being defined exist independently of our definitions. (For instance, 'the tallest boy in the class' appears unproblematic despite being impredicative.) If this is correct, then (iv) would follow. Finally, if mathematics is about some independently existing reality, then every mathematical problem has a unique and determinate answer, which provides at least some motivation for Hilbertian optimism.

The truth of mathematical platonism would therefore have important consequences within mathematics itself. It would justify the classical methods associated with working realism and encourage the search for new axioms to settle questions which are left open by our current mathematical theories.

However, working realism does not in any obvious way imply platonism. Although working realism says that we are justified in using the platonistic language of contemporary mathematics, this falls short of platonism in at least two ways. As the above discussion of truth-value realism showed, the platonistic language of mathematics can be analysed in such a way as to avoid reference to and quantification over mathematical objects. Moreover, even if a face-value analysis of the language of mathematics could be justified, this would support object realism but not platonism. An additional argument would be needed for the third component of platonism, namely, Independence.

4

ARISTOTLE'S METAPHYSICS

4.1 INTRODUCTION

For many people, Plato is the archetypal other-worldly philosopher and Aristotle the archetypal this-worldly philosopher. Plato found reality to lie in things wholly remote from sense, while Aristotle took form to be typically embodied in matter and considered it his job as a philosopher to make sense of the here and now. The contrast is to some extent overdrawn, for Aristotle, too, believed in pure form, and Plato was sufficiently concerned with the here and now to want radical change in human society. It remains true, nevertheless, that Aristotelianism is in essentials a species of immanent metaphysics—a theory that instructs people on how to take the world they know rather than a theory that gives them news of an altogether different world.

The key concepts in Aristotelianism are substance, form and matter, potentiality and actuality, and cause. Whatever happens involves some substance or substances; unless there were substances, in the sense of concrete existents, nothing whatsoever could be real. Substances, however, are not, as the name might suggest, mere parcels of matter; they are intelligible structures, or forms, embodied in matter. That a thing is of a certain kind means that it has a certain form or structure. But the structure as conceived in Aristotelianism is not merely static. Every substance, in this view, not only has a form but is, as it were, striving to attain its natural form; it is seeking to be in actuality what it is potentially, which is in effect to be a proper specimen of its kind. Because this is so, explanation in this system must be given in teleological rather than mechanical terms. For Aristotle, form is the determining element in the universe, but it operates by drawing things on, so that they become what they have it in themselves to be rather than by acting as a constant efficient cause. The notion of an efficient cause has a role in Aristotelianism. As Aristotle put it, it takes a human being, a developed specimen of the kind, to beget a human being. It is, however, a subordinate role and yields pride of place to a different idea—namely, form considered as purpose.

For reasons connected with his astronomy, Aristotle postulated a God. His God, however, had nothing to do with the universe; it was not his creation, and he was, of necessity, indifferent to its vicissitudes. It is a mistake to imagine that everything in the Aristotelian universe is trying to fulfill a purpose that God has ordained for it. On the contrary, the teleology of which use is here made is unconscious; although things all tend to an end, they do not in general consciously seek that end. They are like organs in a living body that fulfill a function and yet seemingly have not been put there for that purpose.

As this last remark will suggest, an important source of Aristotelian thought is reflection on natural growth and decay. Aristotle, who was the son of a doctor, was himself a pioneer in natural history, and it is not surprising that he thought in biological terms. What is surprising, and gives his system a continuing interest, is the extent to which he succeeded in applying ideas in fields that are remote from their origin. He was without doubt more successful in some fields than in others—in dealing with the phenomena of social life, for instance, as opposed to those of physical reality. His results overall, however, were impressive enough for his system not only to dominate Western thought for many centuries but to constitute a challenge even today. People still, in many respects, think like Aristotle, and, as long as that is so, Aristotelianism will remain a live metaphysical option.

4.2 THE SUBJECT MATTER OF ARISTOTLE'S METAPHYSICS

Aristotle himself described his subject matter in a variety of ways: as 'first philosophy', or 'the study of being qua being', or 'wisdom', or 'theology'. A comment on these descriptions will help to clarify Aristotle's topic.

In Metaphysics A.1, Aristotle says that "everyone takes what is called 'wisdom' (sophia) to be concerned with the primary causes (aitia) and the starting-points and it is these causes and principles that he proposes to study in this work. It is his customary practice to begin an inquiry by reviewing the opinions previously held by others, and that is what he does here, as Book A continues with a history of the thought of his predecessors about causes and principles.

These causes and principles are clearly the subject matter of what he calls 'first philosophy'. But this does not mean the branch of philosophy that should be studied first. Rather, it concerns issues that are in some sense the most fundamental or at the highest level of generality. Aristotle distinguished between things that are "better known to us" and things that are "better known in themselves," and maintained that we should begin our study of a given topic with things better known to us and arrive ultimately at an understanding of things better known in themselves. The principles studied by 'first philosophy' may seem very general and abstract, but they are, according to Aristotle, better known in themselves, however remote they may seem from the world of ordinary experience. Still, since they are to be studied only by one who has already studied nature, they are quite appropriately described as coming "after the Physics."

Aristotle's description 'the study of being qua being' is frequently and easily misunderstood, for it seems to suggest that there is a single (albeit special) subject matter—being qua being—that is under investigation. But Aristotle's description does not involve two things—(1) a study and (2) a subject matter (being qua being)—for he did not think that there is any such subject matter as 'being qua being'. Rather, his description involves three things: (1) a study, (2) a subject matter (being), and (3) a manner in which the subject matter is studied (qua being).

Aristotle's Greek word that has been Latinized as 'qua' means roughly 'in so far as' or 'under the aspect'. A study of x qua y, then, is a study of x that concerns itself solely with the y aspect of x. So Aristotle's study does not concern some recondite subject matter known as 'being qua being'. Rather it is a study of being, or better, of beings—of things that can be said to be—that studies them in a particular way: as beings, in so far as they are beings.

Of course, first philosophy is not the only field of inquiry to study beings. Natural science and mathematics also study beings, but in different ways, under different aspects. The natural scientist studies them as things that are subject to the laws of nature, as things that move and undergo change. That is, the natural scientist studies things qua movable (i.e., in so far as they are subject to change). The mathematician studies things qua countable and measurable. The metaphysician, on the other hand, studies them in a more general and abstract way—qua beings. So first philosophy studies the causes and principles of beings qua beings. Aristotle adds that for this reason it studies the causes and principles of substances (ousiai). We will explain this connection in Section 3 below.

In Book E, Aristotle adds another description to the study of the causes and principles of beings qua beings. Whereas natural science studies objects that are material and subject to change, and mathematics studies objects that although not subject to change are nevertheless not separate from matter, there is still room for a science that studies things that are eternal, not subject to change, and independent of matter. Such a science, he says, is theology, and this is the "first" and "highest" science. Aristotle's identification of theology, so conceived, with the study of being qua being has proved challenging to his interpreters. We discuss this identification in Section 14 below.

Finally, we may note that in Book B, Aristotle delineates his subject matter in a different way, by listing the problems or perplexities (aporiai) he hopes to deal with. Characteristic of these perplexities, he says, is that they tie our thinking up in knots. They include the following, among others: Are sensible substances the only ones that exist, or are there others besides them? Is it kinds or individuals that are the elements and principles of things? And if it is kinds, which ones: the most generic or the most specific? Is there a cause apart from matter? Is there anything apart from material compounds? Are the principles limited, either in number or in kind? Are the principles of perishable things themselves perishable? Are the principles universal or particular, and do they exist potentially or actually? Are mathematical objects substances? If they are, are they separate from or do they always belong to sensible things? And ("the hardest and most perplexing of all," Aristotle says) are unity and being the substance of things, or are they attributes of some other subject? In the remainder of Book B, Aristotle presents arguments on both sides of each of these issues, and in subsequent books he takes up many of them again. But it is not always clear precisely how he resolves them, and it is possible that Aristotle did not think that the Metaphysics contains definitive solutions to all of these perplexities.

4.2.1 The Categories

To understand the problems and project of Aristotle's Metaphysics, it is best to begin with one of his earlier works, the Categories. Although placed by long tradition among his logical works, due to its analysis of the terms that make up the propositions out of which deductive inferences are constructed, the Categories begins with a strikingly general and exhaustive account of the things there are (ta onta)—beings. According to this account, beings can be divided into ten distinct categories. (Although Aristotle never says so, it is tempting to suppose that these categories are mutually exclusive and jointly exhaustive of the things there are.) They include substance, quality, quantity, and relation, among others. Of these categories of beings, it is the first, substance (ousia), to which Aristotle gives a privileged position.

Substances are unique in being independent things; the items in the other categories all depend somehow on substances. That is, qualities are the qualities of substances; quantities are the amounts and sizes that substances come in; relations are the way substances stand to one another. These various non-substances all owe their existence to substances—each of them, as Aristotle puts it, exists only 'in' a subject. That is, each non-substance "is in something, not as a part, and cannot exist separately from what it is in". Indeed, it becomes clear that substances are the subjects that these ontologically dependent non-substances are 'in'.

Each member of a non-substance category thus stands in this inherence relation (as it is frequently called) to some substance or other—color is always found in bodies, knowledge in the soul. Neither whiteness nor a piece of grammatical knowledge, for example, is capable of existing on its own. Each requires for its existence that there be some substance in which it inheres.

In addition to this fundamental inherence relation across categories, Aristotle also points out another fundamental relation that obtains between items within a single category. He describes this as the relation of "being said of a subject," and his examples make clear that it is the relation of a more general to a less general thing within a single category. Thus, man is 'said of' a particular man, and animal is 'said of' man, and therefore, as Aristotle points out, animal is 'said of' the particular man also. The 'said of' relation, that is to say, is transitive. So the genus is 'said of' the species and both genus and species are 'said of' the particular. The same holds in non-substance categories. In the category of quality, for example, the genus (color) is 'said of' the species (white) and both genus and species are 'said of' the particular white. There has been considerable scholarly dispute about these particulars in nonsubstance categories.

The language of this contrast ('in' a subject vs. 'said of' a subject) is peculiar to the Categories, but the idea seems to recur in other works as the distinction between accidental vs. essential predication. Similarly, in works other than the Categories, Aristotle uses the label 'universals' for the things that are "said of many;" things that are not universal he calls 'particulars'. Although he does not use these labels in the Categories, it is not misleading to say that the doctrine of the Categories is that each category contains a hierarchy of universals and particulars, with each universal being 'said of' the lower-level universals and particulars that fall beneath it. Each category thus has the structure of an upside-down tree. At the top of the tree are the most generic items in that category.

The individuals in the category of substance play a special role in this scheme. Aristotle calls them "primary substances" for without them, as he says, nothing else would exist. Indeed, Aristotle offers an argument to establish the primary substances as the fundamental entities in this ontology. Everything that is not a primary substance, he points out, stands in one of the two relations to primary substances. A genus, such as animal, is 'said of' the species below it and, since they are 'said of' primary substances, so is the genus. Thus, everything in the category of substance that is not itself a primary substance is, ultimately, 'said of' primary substances. And if there were no primary substances, there would be no "secondary" substances, either. For these secondary substances are just the ways in which the primary substances are fundamentally classified within the category of substance. As for the members of non-substance categories, they too depend for their existence on primary substances. A universal in a non-substance category, e.g., color, in the category of quality, is 'in' body, Aristotle tells us, and therefore in individual bodies. For color could not be 'in' body, in general, unless it were 'in' at least some particular bodies. Similarly, particulars in non-substance categories cannot exist on their own. E.g., a determinate shade of color, or a particular and non-shareable bit of that shade, is not capable of existing on its own—if it were not 'in' at least some primary substance, it would not exist. So primary substances are the basic entities—the basic "things that there are"—in the world of the Categories.

4.2.2 The Role of Substance in the Study of Being Qua Being

The Categories leads us to expect that the study of being in general will crucially involve the study of substance, and when we turn to the Metaphysics we are not disappointed. First, in Metaphysics Γ Aristotle argues in a new way for the ontological priority of substance; and then, in Books Z, H, and Θ, he wrestles with the problem of what it is to be a substance. We will begin with Γ's account of the central place of substance in the study of being qua being.

As we noted above, metaphysics (or, first philosophy) is the science which studies being qua being. In this respect it is unlike the specialized or departmental sciences, which study only part of being (only some of the things that exist) or study beings only in a specialized way.

Consider an analogy. There are dining tables, and there are tide tables. A dining table is a table in the sense of a smooth flat slab fixed on legs; a tide table is a table in the sense of a systematic arrangement of data in rows and columns. But there is not a single sense of 'table' which applies to both the piece of furniture at which I am writing these words and to the small booklet that lies upon it. Hence it would be foolish to expect that there is a single science of tables, in general, that would include among its objects both dining tables and tide tables. Tables, that is to say, do not constitute a single kind with a single definition, so no single science, or field of knowledge, can encompass precisely those things that are correctly called 'tables'.

If the term 'being' were ambiguous in the way that 'table' is, Aristotle's science of being qua being would be as impossible as a science of tables qua tables. But, Aristotle argues in Γ.2, 'being' is not ambiguous in this way. 'Being', he tells us, is 'said in many ways' but it is not merely 'homonymous', i.e., sheerly ambiguous. Rather, the various senses of 'being' have what he calls a 'pros hen' ambiguity—they are all related to a single central sense.

Aristotle explains his point by means of some examples that he takes to be analogous to 'being'. Consider the terms 'healthy' and 'medical'. Neither of these has a single definition that applies uniformly to all cases: not every healthy (or medical) thing is healthy (medical) in the same sense of 'healthy' ('medical'). There is a range of things that can be called 'healthy': people, diets, exercise, complexions, etc. Not all of these are healthy in the same sense. Exercise is healthy in the sense of being productive of health; a clear complexion is healthy in the sense of being symptomatic of health; a person is healthy in the sense of having good health.

But notice that these various senses have something in common: a reference to one central thing, health, which is actually possessed by only some of the things that are spoken of as 'healthy', namely, healthy organisms, and these are said to be healthy in the primary sense of the term. Other things are considered healthy only in so far as they are appropriately related to things that are healthy in this primary sense.

The situation is the same, Aristotle claims, with the term 'being'. It, too, has a primary sense as well as related senses in which it applies to other things because they are appropriately related to things that are called 'beings' in the primary sense. The beings in the primary sense are substances; the beings in other senses are the qualities, quantities, etc., that belong to substances. An animal, e.g., a horse, is a being, and so is a color, e.g, white, a being. But a horse is a being in the primary sense—it is a substance—whereas the color white (a quality) is a being only because it qualifies some substance. An account of the being of anything that is, therefore, will ultimately have to make some reference to substance. Hence, the science of being qua being will involve an account of the central case of beings—substances.

The Fundamental Principles: Axioms

Before embarking on this study of substance, however, Aristotle goes on in Book Γ to argue that first philosophy, the most general of the sciences, must also address the most fundamental principles—the common axioms—that are used in all reasoning. Thus, first philosophy must also concern itself with the principle of non-contradiction (PNC): the principle that "the same thing cannot at the same time belong and also not belong to the same thing and in the same respect". This, Aristotle says, is the most certain of all principles, and it is not just a hypothesis. It cannot, however, be proved, since it is employed, implicitly, in all proofs, no matter what the subject matter. It is a first principle, and hence is not derived from anything more basic.

What, then, can the science of first philosophy say about the PNC? It cannot offer a proof of the PNC, since the PNC is presupposed by any proof one might offer—any purported proof of the PNC would therefore be circular. Aristotle thus does not attempt to prove the PNC; in the subsequent chapters of Γ he argues, instead, that it is impossible to disbelieve the PNC. Those who would claim to deny the PNC cannot, if they have any beliefs at all, believe that it is false. For one who has a belief must, if he is to express this belief to himself or to others, say something—he must make an assertion. He must, as Aristotle says, signify something. But the very act of signifying something is possible only if the PNC is accepted. Without accepting the PNC, one would have no reason to think that his words have any signification at all—they could not mean one thing rather than another. So anyone who makes any assertion has already committed himself to the PNC. Aristotle thus does not argue that the PNC is a necessary truth (that is, he does not try to prove the PNC); rather, he argues that the PNC is indubitable.

What is Substance?

Aristotle takes up the promised study of substance. He begins by reiterating and refining some of what he said in that 'being' is said in many ways, and that the primary sense of 'being' is the sense in which substances are beings. Here, however, he explicitly links the secondary senses of 'being' to the non-substance categories. The primacy of substance leads Aristotle to say that the age-old question 'What is being?' "is just the question 'What is substance?'".

One might have thought that this question had already been answered in the Categories. There we were given, as examples of primary substances, an individual man or horse, and we learned that a primary substance is "what is neither in a subject nor said of a subject". This would seem to provide us with both examples of, and criteria for being, primary substances. But in Metaphysics Z, Aristotle does not seem to take either the examples or the criteria for granted.

In Z.2 he recounts the various answers that have been given to the question of which things are substances—bodies, things more basic than bodies (surfaces, lines, and points), imperceptible things (such as Platonic Forms and mathematical objects)—and seems to regard them all as viable candidates at this point. He does not seem to doubt that the clearest examples of substances are perceptible ones, but leaves open the question whether there are others as well.

Before answering this question about examples, however, he says that we must first answer the question about criteria: what is it to be a substance? The negative criterion of the Categories tells us only which things are substances. But even if we know that something is a substance, we must still say what makes it a substance—what the cause is of its being a substance. This is the question to which Aristotle next turns. To answer it is to identify, as Aristotle puts it, the substance of that thing.

Substance, Matter, and Subject

Z.3 begins with a list of four possible candidates for being the substance of something: essence, universal, genus, and subject. Presumably, this means that if x is a substance, then the substance of x might be either (i) the essence of x, or (ii) some universal predicated of x, or (iii) a genus that x belongs to, or (iv) a subject of which x is predicated. The first three candidates are taken up in later chapters, and Z.3 is devoted to an examination of the fourth candidate: the idea that the substance of something is a subject of which it is predicated.

A subject, Aristotle tells us, is "that of which the other things are said, but which is itself not further said of any other thing". This characterization of a subject is reminiscent of the language of the Categories, which tells us that a primary substance is not predicated of anything else, whereas other things are predicated of it. Candidate thus seems to reiterate the Categories criterion for being a substance. But there are two reasons to be wary of drawing this conclusion. First, whereas the subject criterion of the Categories told us that substances were the ultimate subjects of predication, the subject criterion envisaged here is supposed to tell us what the substance of something is. So what it would tell us is that if x is a substance, then the substance of x—that which makes x a substance—is a subject that x is predicated of. Second, as his next comment makes clear, Aristotle has in mind something other than this Categories idea. For the subject that he here envisages, he says, is either matter or form or the compound of matter and form. These are concepts from Aristotle's Physics, and none of them figured in the ontology of the Categories. To appreciate the issues Aristotle is raising here, we must briefly compare his treatment of the notion of a subject in the Physics with that in the Categories.

In the Categories, Aristotle was concerned with subjects of predication: what are the things we talk about, and ascribe properties to? In the Physics, his concern is with subjects of change: what is it that bears contrary predicates and persists through a process of change? But there is an obvious connection between these conceptions of a subject, since a subject of change must have one predicate belonging to it at one time that does not belong to it at another time. Subjects of change, that is, are also subjects of predication.

In the Categories, individual substances (a man, a horse) were treated as fundamental subjects of predication. They were also understood, indirectly, as subjects of change. "A substance, one and the same in number, can receive contraries. An individual man, for example, being one and the same, becomes now pale and now dark, now hot and now cold, now bad and now good". These are changes in which substances move, or alter, or grow. What the Categories did not explore, however, are changes in which substances are generated or destroyed. But the theory of change Aristotle develops in the Physics requires some other subject for changes such as these—a subject of which substance is predicated—and it identifies matter as the fundamental subject of change. Change is seen in the Physics as a process in which matter either takes on or loses form.

The concepts of matter and form, as we noted, are absent from the Categories. Individual substances—this man or that horse—apart from their accidental characteristics—the qualities, etc., that inhere in them—are viewed in that work as essentially simple, unanalyzable atoms. Although there is metaphysical structure to the fact that, e.g., this horse is white, the fact that this is a horse is a kind of brute fact, devoid of metaphysical structure. This horse is a primary substance, and horse, the species to which it belongs, is a secondary substance. But there is no predicative complex corresponding to the fact that this is a horse in the way that there is such a complex corresponding to the fact that this horse is white.

But from the point of view of the Physics, substantial individuals are seen as predicative complexes; they are hylomorphic compounds—compounds of matter and form—and the subject criterion looks rather different from the hylomorphic perspective. Metaphysics Z.3 examines the subject criterion from this perspective.

Matter, form, and the compound of matter and form may all be considered subjects, Aristotle tells us, but which of them is substance? The subject criterion by itself leads to the answer that the substance of x is an entirely indeterminate matter of which x is composed. For form is predicated of matter as subject, and one can always analyze a hylomorphic compound into its predicates and the subject of which they are predicated. And when all predicates have been removed, the subject that remains is nothing at all in its own right—an entity all of whose properties are accidental to it. The resulting subject is matter from which all form has been expunged. So the subject criterion leads to the answer that the substance of x is the formless matter of which it is ultimately composed.

But Aristotle rejects this answer as impossible, claiming that substance must be "separable" and "this something", and implying that matter fails to meet this requirement. Precisely what the requirement amounts to is a matter of considerable scholarly debate, however. A plausible interpretation runs as follows. Separability has to do with being able to exist independently (x is separable from y if x is capable of existing independently of y), and being a this something means being a determinate individual. So a substance must be a determinate individual that is capable of existing on its own. (One might even hold, although this is controversial, that on Aristotle's account not every "this something" is also "separable." A particular color or shape might be considered a determinate individual that is not capable of existing on its own—it is always the color of shape of some substance or other.) But matter fails to be simultaneously both chôriston and tode ti. The matter of which a substance is composed may exist independently of that substance (think of the wood of which a desk is composed, which existed before the desk was made and may survive the disassembly of the desk), but it is not as such any definite individual—it is just a quantity of a certain kind of matter. Of course, the matter may be construed as constituting a definite individual substance (the wood just is, one might say, the particular desk it composes), but it is in that sense not separable from the form or shape that makes it that substance (the wood cannot be that particular desk unless it is a desk). So although matter is in a sense separable and in a sense a this something, it cannot be both separable and a this something. It thus does not qualify as the substance of the thing whose matter it is.

Substance and Essence

Aristotle turns in Z.4 to a consideration of the next candidate for substance: essence. ('Essence' is the standard English translation of Aristotle's curious phrase to ti ên einai, literally "the what it was to be" for a thing. This phrase so boggled his Roman translators that they coined the word essentia to render the entire phrase, and it is from this Latin word that ours derives. Aristotle also sometimes uses the shorter phrase to ti esti, literally "the what it is," for approximately the same idea.) In his logical works, Aristotle links the notion of essence to that of definition (horismos)—"a definition is an account (logos) that signifies an essence"—and he links both of these notions to a certain kind of per se predication (kath' hauto, literally, "in respect of itself," or "intrinsically")—"what belongs to a thing in respect of itself belongs to it in its essence (en tôi ti esti)" for we refer to it "in the account that states the essence". He reiterates these ideas in Z.4: "there will be an essence only of those things whose logos is a definition", "the essence of each thing is what it is said to be intrinsically". It is important to remember that for Aristotle, one defines things, not words. The definition of tiger does not tell us the meaning of the word 'tiger'; it tells us what it is to be a tiger, what a tiger is said to be intrinsically. Thus, the definition of tiger states the essence—the "what it is to be" of a tiger, what is predicated of the tiger per se.

Aristotle's preliminary answer to the question "What is substance?" is that substance is essence, but there are important qualifications. For, as he points out, "definition (horismos), like 'what-it-is' (ti esti), is said in many ways too". That is, items in all the categories are definable, so items in all the categories have essences—just as there is an essence of man, there is also an essence of white and an essence of musical. But, because of the pros hen equivocity of 'is', such essences are secondary—"in the primary (protôs) and unconditional way (haplôs) definition and the essence belong to substances". Thus, Z.4 tells us, it is only these primary essences that are substances. Aristotle does not here work out the details of this "hierarchy of essences".

Aristotle goes on to argue that if something is "primary" and "spoken of in respect of itself (kath' hauto legomenon)" it is one and the same as its essence. The precise meaning of this claim, as well as the nature and validity of the arguments offered in support of it, are matters of scholarly controversy. But it does seem safe to say that Aristotle thinks that an "accidental unity" such as a pale man is not a kath' hauto legomenon (since pallor is an accidental characteristic of a man) and so is not the same as its essence. Pale man, that is to say, does not specify the "what it is" of any primary being, and so cannot be an essence of the primary kind. As Z.4 has already told us, essence, in the primary sense, "will belong to things that are species of a genus and to nothing else". Man is a species, and so there is an essence of man; but pale man is not a species and so, even if there is such a thing as the essence of pale man, it is not, at any rate, a primary essence.

At this point there appears to be a close connection between the essence of a substance and its species (eidos), and this might tempt one to suppose that Aristotle is identifying the substance of a thing (since the substance of a thing is its essence) with its species. (A consequence of this idea would be that Aristotle is radically altering his conception of the importance of the species, which in the Categories he called a secondary substance, that is, a substance only in a secondary sense.) But such an identification would be a mistake, for two reasons. First, Aristotle's point at 1030a11 is not that a species is an essence, but that an essence of the primary kind corresponds to a species (e.g., man) and not to some more narrowly delineated kind (e.g., pale man). Second, the word 'eidos', which meant 'species' in the logical works, has acquired a new meaning in a hylomorphic context, where it means 'form' (contrasted with 'matter') rather than 'species' (contrasted with 'genus'). In the conceptual framework of Metaphysics Z, a universal such as man or horse—which was called a species and a secondary substance in the Categories—is construed as "not substance but rather a compound of a sort, of this account and this matter taken universally". The eidos that is primary substance in Book Z is not the species that an individual substance belongs to but the form that is predicated of the matter of which it is composed.

Substances as Hylomorphic Compounds

The role of form in this hylomorphic context is the topic of Z.7–9. (Although these chapters were almost certainly not originally included in Book Z—there is no reference to them, for example, in the summary of Z given in H.1, which skips directly from Z.6 to Z.10—they provide a link between substance and form and thus fill what would otherwise be a gap in the argument.) Since individual substances are seen as hylomorphic compounds, the role of matter and form in their generation must be accounted for. Whether we are thinking of natural objects, such as plants and animals, or artifacts, such as houses, the requirements for generation are the same. We do not produce the matter (to suppose that we do leads to an infinite regress) nor do we produce the form (what could we make it out of?); rather, we put the form into the matter, and produce the compound. Both the matter and the form must pre-exist. But the source of motion in both cases—what Aristotle calls the "moving cause" of the coming to be—is the form.

In production that results from craft (or art, technê), "the form is in the soul" of the craftsman. For example, "the craft is the form" and the craft, i.e., the form, is in the understanding, and hence in the soul, of the builder. The builder has in mind the plan or design for a house and he knows how to build; he then "enmatters" that plan or design by putting it into the materials out of which he builds the house. In natural production, the form is found in the parent, where "the begetter is of this same sort as the begotten (not that they are the same thing, certainly, nor one in number, but one in form)—for example, in the case of natural things. For human begets human". But in either case, the form pre-exists and is not produced.

As for what is produced in such hylomorphic productions, it is correctly described by the name of its form, not by that of its matter. What is produced is a house or a man, not bricks or flesh. Of course, what is made of gold may still be described in terms of its material components, but we should call it not "gold" but "golden". For if gold is the matter out of which a statue is made, there was gold present at the start, and so it was not gold that came into being. It was a statue that came into being, and although the statue is golden—i.e., made of gold—it cannot be identified with the gold of which it was made.

The essence of such a hylomorphic compound is evidently its form, not its matter. As Aristotle says "by form I mean the essence of each thing and the primary substance", and "by the substance without matter I mean the essence". It is the form of a substance that makes it the kind of thing that it is, and hence it is form that satisfies the condition initially required for being the substance of something. The substance of a thing is its form.

Substance and Definition

In Z.10 and 11, Aristotle returns to the consideration of essence and definition left off in Z.6, but now within the hylomorphic context developed in Z.7–9. The main question these chapters consider is whether the definition of x ever includes a reference to the matter of x. If some definitions include a reference to matter, then the link between essence and form would seem to be weakened.

Aristotle begins Z.10 by endorsing the following principle about definitions and their parts: "a definition is an account, and every account has parts, and as the account is to the thing, so the part of the account is to the part of the thing". That is, if y is a part of a definable thing x, then the definition of x will include as a part something z that corresponds to y. Indeed, z must stand to y in the same relation that the definition of x stands in to x; that is, z is the definition of y. So, according to this principle, the definition of a thing will include the definitions of its parts.

In a way, this consequence of the principle seems very plausible, given Aristotle's idea that it is universals that are definable. Consider as a definiendum a universal, such as man, and its definiens, rational animal. The parts of this definiens are the universals rational and animal. If these parts are, in turn, definable, then each should be replaced, in the definition of man, with its own definition, and so on. In this way the complete and adequate definition of a universal such as man will contain no parts that are further definable. All proper, or completely analyzed, definitions are ultimately composed of simple terms that are not further definable.

But the implication of this idea for the definitions of hylomorphic compounds is obvious: since matter appears to be a part of such a compound, the definition of the compound will include, as a part, the definitions of its material components. And this consequence seems implausible to Aristotle. A circle, for example, seems to be composed of two semicircles (for it obviously may be divided into two semicircles), but the definition of circle cannot be composed of the definitions of its two semicircular parts. For, as Aristotle points out, semicircle is defined in terms of circle, and not the other way around. His point is well taken, for if circles were defined in terms of semicircles, then presumably semicircles would be defined in terms of the quarter-circles of which they are composed, and so on, ad infinitum. The resulting infinite regress would make it impossible to define circle at all, for one would never reach the ultimate "simple" parts of which such a definition would be composed.

Aristotle flirts with the idea of distinguishing between different senses in which one thing can be a part of another, but instead proposes a different solution: to specify carefully the whole of which the matter is allegedly a part. "And of the compound statue the bronze is a part, but of what is said to be a statue as form it is not a part". Similarly, "even if the line, when divided, passes away into the halves, or the human into the bones, sinews, and flesh, it is not the case that because of this they are composed of these as being parts of the substance". Rather, "what is divided into these as into matter is not the substance but the compound".

In restating his point "yet more perspicuously", Aristotle notes parenthetically another important aspect of his theory of substance. He reiterates the priority of form, and its parts, to the matter into which a compound is divided, and notes that "the soul of animals is the substance that is in accord with the account and is the form and the essence". The idea recurs in where he announces that "it is clear too that the soul is primary substance, whereas the body is matter". It is further developed, in the Metaphysics, as we will and especially in De Anima. For more detail on this topic, see Section 3 of the entry on Aristotle's psychology.

Returning now to the problem raised by the apparent need to refer to matter in the definition of a substance, we may note that the solution Aristotle offered in Z.10 is only partially successful. His point seems to be that whereas bronze may be a part of a particular statue, neither that particular batch of bronze nor even bronze in general enters into the essence of statue, since being made of bronze is no part of what it is to be a statue. But that is only because statues, although they must be made of some kind of matter, do not require any particular kind of matter. But what about kinds of substances that do require particular kinds of matter? Aristotle's distinction between form and compound cannot be used in such cases to isolate essence from matter. Thus there may after all be reasons for thinking that reference to matter will have to intrude into at least some definitions.

In Z.11, Aristotle addresses just such a case. "The form of the human is always found in flesh and bones and parts of this sort," Aristotle writes. The point is not just that each particular man must be made of matter, but that each one must be made of matter of a particular kind—flesh and bones, etc. "Some things," he continues, "presumably are this in this"), i.e., a particular form in a particular kind of matter, so that it is not possible to define them without reference to their material parts. Nevertheless, Aristotle ends Z.11 as if he has defended the claim that definition is of the form alone. Perhaps his point is that whenever it is essential to a substance that it be made of a certain kind of matter (e.g., that man be made of flesh and bones, and that one "could not make a saw of wool or wood,") this is in some sense a formal or structural requirement. A kind of matter, after all, can itself be analyzed hylomorphically—bronze, for example, is a mixture of copper and tin according to a certain ratio or formula (logos), which is in turn predicated of some more generic underlying subject. The reference to matter in a definition will thus always be to a certain kind of matter, and hence to a predicate, rather than a subject. At any rate, if by 'matter' one has in mind the ultimate subject alluded to in Z.3 (so-called 'prime matter'), there will be no reference to it in any definition, "for it is indefinite".

Z.12 introduces a new problem about definitions—the so-called "unity of definition." The problem is this: definitions are complex, so what accounts for the definiendum being one thing, rather than many? Suppose that man is defined as two-footed animal; "why, then, is this one and not many—animal and two-footed?". Presumably, Aristotle has in mind his discussion in Z.4 of such "accidental unities" as a pale man. The difference cannot be that our language contains a single word for a two-footed animal, but no single word for a pale man, for Aristotle has already conceded that we might very well have had a single term for a pale man, but that would still not make the formula 'pale man' a definition nor pale man an essence.

Aristotle proposes a solution that applies to definitions reached by the "method of division." According to this method, one begins with the broadest genus containing the species to be defined, and divides the genus into two sub-genera by means of some differentia. One then locates the definiendum in one of the sub-genera, and proceeds to divide this by another differentia, and so on, until one arrives at the definiendum species. This is a classic definition by genus and differentia. Aristotle's proposal is that "the division should take the differentia of the differentia" (1038a9). For example, if one uses the differentia footed to divide the genus animal, one then uses a differentia such as cloven-footed for the next division. If one divides in this way, Aristotle claims, "it is evident that the ultimate differentia will be the substance of the thing and its definition". For each "differentia of a differentia" entails its predecessor, and so the long chain of differentiae can be replaced simply by the ultimate differentia, since it entails all of its predecessors. As Aristotle points out, it would be redundant to include any of the differentiae in the chain other than the ultimate one: "when we say footed two-footed animal ... we shall be saying the same thing several times over".

This proposal shows how a long string of differentiae in a definition can be reduced to one, but it does not solve the problem of the unity of definition. For we are still faced with the apparent fact that genus + differentia constitutes a plurality even if the differentia is the ultimate, or "completing," one. It is not surprising, then, that Aristotle returns to the problem of unity later and offers a different solution.

Substances and Universals

At this point, we seem to have a clear idea about the nature of substantial form as Aristotle conceives of it. A substantial form is the essence of a substance, and it corresponds to a species. Since it is an essence, a substantial form is what is denoted by the definiens of a definition. Since only universals are definable, substantial forms are universals. That substantial forms are universals is confirmed by Aristotle's comment, at the end that "Socrates and Callias ... are distinct because of their matter ... but the same in form". For them to be the same in form is for them to have the same form, i.e., for one and the same substantial form to be predicated of two different clumps of matter. And being "predicated of many" is what makes something a universal.

But Z.13 throws our entire understanding into disarray. Aristotle begins by returning to the candidates for the title of ousia introduced in Z.3, and points out that having now discussed the claims of the subject and the essence, it is time to consider the third candidate, the universal. But the remainder of the chapter consists of a barrage of arguments to the conclusion that universals are not substances.

Z.13 therefore produces a fundamental tension in Aristotle's metaphysics that has fragmented his interpreters. Some maintain that Aristotle's theory is ultimately inconsistent, on the grounds that it is committed to all three of the following propositions:

(i) Substance is form.

(ii) Form is universal.

(iii) No universal is a substance.

Others have provided interpretations according to which Aristotle does not maintain all of (i)–(iii), and there is a considerable variety of such interpretations, too many to be canvassed here. But there are two main, and opposed, lines of interpretation. According to one, Aristotle's substantial forms are not universals after all, but each belongs exclusively to the particular whose form it is, and there are therefore as many substantial forms of a given kind as there are particulars of that kind. According to the other, Aristotle's arguments in Z.13 are not intended to show that no universal is a substance, tout court, but some weaker thesis that is compatible with there being only one substantial form for all of the particulars belonging to the same species.

It would be foolish to attempt to resolve this issue within the confines of the present entry, as it is perhaps the largest, and most disputed, single interpretative issue concerning Aristotle's Metaphysics. we will, instead, mention some of the main considerations brought up on each side of this dispute, and give our reasons for thinking that substantial forms are universals.

The idea that substantial forms are particulars is supported by Aristotle's claims that a substance is "separable and this something " On the other side, the idea that substantial forms are universals is supported by Aristotle's claims that substances are, par excellence, the definable entities, that definition is of the universal, and that it is impossible to define particulars.

In our opinion, the indefinability of particulars makes it impossible for substantial forms to be particulars. If there were a substantial form that is unique to some sensible particular, say Callias, then the definition corresponding to that form, or essence, would apply uniquely to Callias—it would define him, which is precisely what Aristotle says cannot be done. The question, then, is whether the evidence against substantial forms being universals can be countered. This is less clear, but the following considerations are relevant. (1) Aristotle's claim that a substantial form is an individual does not exclude its being a universal (katholou). Universals are contrasted with particulars (kath' hekasta), not individuals (although Aristotle does sometimes ignore the distinction between tode ti and kath' hekaston). What makes something a tode ti is its being a fully determinate thing, not further differentiable; what makes something a kath' hekaston is its being a particular thing, unrepeatable, and not predicated of anything else. There is thus the possibility of a universal tode ti—a fully determinate universal not further divisible into lower-level universals, but predicated of numerous particulars. (2) The claim that there are no universals apart from particulars needs to be understood in context. When Aristotle asserts that "there is not some animal ... beyond the particular ones (ta tina)" he is just as likely to be referring to the particular kinds of animals as he is to particular specimens. If so, his point may be that a generic kind, such as animal, is ontologically dependent on its species, and hence on the substantial forms that are the essences of those species. (3) The arguments against the substantiality of universals are presented as part of a give-

and-take investigation of the perplexities involved in the notion of substantial form. It is not clear, therefore, whether the blanket claim "No universal is a substance" is intended to be accepted without qualification. Indeed, a closer examination of the arguments may show that qualifications are required if the arguments are to be cogent. For example, the argument is based on the premise that the substance of x is peculiar (idion) to x. It then draws the conclusion that a universal cannot be the substance of all of its instances (for it could not be idion to all of them), and concludes that it must be the substance "of none." But note that this conclusion does not say that no universal can be a substance, but only that no universal can be the substance of any of its instances. Aristotle's point may be that since form is predicated of matter, a substantial form is predicated of various clumps of matter. But it is not the substance of those clumps of matter, for it is predicated accidentally of them. The thing with which it is uniquely correlated, and of which it is the substance, is not one of its instances, but is the substantial form itself. This conclusion should not be surprising in light of Aristotle's claim that "each substance is one and the same as its essence." A universal substantial form just is that essence.

Substance as Cause of Being

Although Aristotle is careful to distinguish four different kinds of cause (or four different senses of 'cause'), it is important to note that he claims that one and the same thing can be a cause in more than one sense. As he puts it, "form, mover, and telos often coincide". And in De Anima he is perfectly explicit that the soul, which is the form or essence of a living thing, "is a cause in three of the ways we have distinguished"—efficient, formal, and final.

Let us return to Aristotle's discussion in Z.17. The job of a cause or principle of being, he notes, is to explain why one thing belongs to another (1041a11); that is, it is to explain some predicational fact. What needs to be explained, for example, is why this is a man, or that is a house. But what kind of a question is this? The only thing that can be a man is a man; the only thing that can be a house is a house. So we would appear to be asking why a man is a man, or why a house is a house, and these seem to be foolish questions that all have the same answer: because each thing is itself (1041a17–20). The questions must therefore be rephrased by taking advantage of the possibility of a hylomorphic analysis. We must ask, e.g., "why are these—for example, brick and stones—a house?" (1041a26). The answer Aristotle proposes is that the cause of being of a substance (e.g., of a house) is the form or essence that is predicated of the matter (e.g., of the bricks and stones) that constitute that substance.

The essence is not always just a formal cause; in some cases, Aristotle says, it is also a final cause (he gives the examples of a house and a bed), and in some cases an efficient cause (1041a29–30). But in any case "what is being looked for is the cause in virtue of which the matter is something—and this is the substance" (1041b6–9) and "the primary cause of its being" (1041b27).

Notice that the explanandum in these cases ("why is this a man?" or "why is that a house?") involves a species predication ("Callias is a man," "Fallingwater is a house"). But the answer Aristotle proposes invokes a hylomorphic analysis of these questions, in which form is predicated of matter. So Callias is a man because the form or essence of man is present in the flesh and bones that constitute the body of Callias; Fallingwater is a house because the form of house is present in the materials of which Fallingwater is made. In general, a species predication is explained in terms of an underlying form predication, whose subject is not the particular compound but its matter. Form predications are thus more basic than their corresponding species predications. A substantial form, as a primary definable, is its own substance, for it is essentially predicated of itself alone. But the substantial form of a material compound, because it is predicated (accidentally) of the matter of the compound, is the cause of the compound's being the kind of thing that it is. The form is therefore, in a derivative way, the substance of the compound, as well.

Actuality and Potentiality

In Metaphysics Z, Aristotle introduces the distinction between matter and form synchronically, applying it to an individual substance at a particular time. The matter of a substance is the stuff it is composed of; the form is the way that stuff is put together so that the whole it constitutes can perform its characteristic functions. But soon he begins to apply the distinction diachronically, across time. This connects the matter/form distinction to another key Aristotelian distinction, that between potentiality (dunamis) and actuality (entelecheia) or activity (energeia).

Aristotle distinguishes between two different senses of the term dunamis. In the strictest sense, a dunamis is the power that a thing has to produce a change. A thing has a dunamis in this sense when it has within it "a starting-point of change in another thing or in itself insofar as it is other". The exercise of such a power is a kinêsis—a movement or process. So, for example, the housebuilder's craft is a power whose exercise is the process of housebuilding. But there is a second sense of dunamis—and it is the one in which Aristotle is mainly interested—that might be better translated as 'potentiality'. For, as Aristotle tells us, in this sense dunamis is related not to movement (kinêsis) but to activity. A dunamis in this sense is not a thing's power to produce a change but rather its capacity to be in a different and more completed state. Aristotle thinks that potentiality so understood is indefinable, claiming that the general idea can be grasped from a consideration of cases. Activity is to potentiality, Aristotle tells us, as "what is awake is in relation to what is asleep, and what is seeing is in relation to what has its eyes closed but has sight, and what has been shaped out of the matter is in relation to the matter".

This last illustration is particularly illuminating. Consider, for example, a piece of wood, which can be carved or shaped into a table or into a bowl. In Aristotle's terminology, the wood has (at least) two different potentialities, since it is potentially a table and also potentially a bowl. The matter (in this case, wood) is linked with potentialty; the substance (in this case, the table or the bowl) is linked with actuality. The as yet uncarved wood is only potentially a table, and so it might seem that once it is carved the wood is actually a table. Perhaps this is what Aristotle means, but it is possible that he does not wish to consider the wood to be a table. His idea might be that not only can a piece of raw wood in the carpenter's workshop be considered a potential table, but the wood composing the completed table is also, in a sense, a potential table. The idea here is that it is not the wood qua wood that is actually a table, but the wood qua table. Considered as matter, it remains only potentially the thing that it is the matter of.

Since Aristotle gives form priority over matter, we would expect him similarly to give actuality priority over potentiality. And that is exactly what we find. Aristotle distinguishes between priority in logos, in time, and in substance. (1) Actuality is prior in logos since we must cite the actuality when we give an account of its corresponding potentiality. Thus, 'visible' means 'capable of being seen'; 'buildable' means 'capable of being built'. (2) As regards temporal priority, by contrast, potentiality may well seem to be prior to actuality, since the wood precedes the table that is built from it, and the acorn precedes the oak that it grows into. Nevertheless, Aristotle finds that even temporally there is a sense in which actuality is prior to potentiality: "the active that is the same in form, though not in number, is prior". A particular acorn is, of course, temporally prior to the particular oak tree that it grows into, but it is preceded in time by the actual oak tree that produced it, with which it is identical in species. The seed must have been preceded by an adult. So in this sense actuality is prior even in time.

(3) Aristotle argues for the priority in substance of actuality over potentiality in two ways. (a) The first argument makes use of his notion of final causality. Things that come to be move toward an end (telos)—the boy becomes a man, the acorn becomes an oak—and "the activity is the end, and it is for the sake of this that the capacity [or potentiality] is acquired. For animals do not see in order that they may have sight, rather they have sight in order that they may see ... matter is potentially something because it may come in the form of it—at any rate, when it is actively something, then it is in the form of it". Form or actuality is the end toward which natural processes are directed. Actuality is therefore a cause in more than one sense of a thing's realizing its potential. As we noted in Section 11, one and the same thing may be the final, formal, and efficient cause of another. Suppose an acorn realizes its potential to become an oak tree. The efficient cause here is the actual oak tree that produced the acorn; the formal cause is the logos defining that actuality; the final cause is the telos toward which the acorn develops—an actual (mature) oak tree.

(b) Aristotle also offers an "even stricter" argument for his claim that actuality is prior in substance to potentiality. A potentiality is for either of a pair of opposites; so anything that is capable of being is also capable of not being. What is capable of not being might possibly not be, and what might possibly not be is perishable. Hence anything with the mere potentiality to be is perishable. What is eternal is imperishable, and so nothing that is eternal can exist only potentially—what is eternal must be fully actual. But the eternal is prior in substance to the perishable. For the eternal can exist without the perishable, but not conversely, and that is what priority in substance amounts to. So what is actual is prior in substance to what is potential.

Unity Reconsidered

In H.6, Aristotle returns to the problem of the unity of definition and offers a new solution based on the concepts of potentiality and actuality. He begins by pointing out that the things whose unity he is trying to explain are those "that have several parts and where the totality of them is not like a heap, but the whole is something beyond the parts". His task is to explain the unity of such complexes.

The problem is insoluble, he says, unless one realizes that "there is on the one hand matter and on the other shape, and the one is potentially and the other actively." Once one realizes this, "then what we are inquiring into will no longer seem to be a puzzle". He offers the following example. Suppose round bronze were the definition of 'cloak'. If someone were to ask "what makes a cloak one thing, a unity?" the answer would be obvious. For bronze is the matter, and roundness is the form. The bronze is potentially round, and round is what the bronze actually is when it has received this form. The cause of the unity of the cloak is just the cause of bronze being made round. Since the cloak is something that was produced, or brought into being, there is no cause of its unity other than the agent who put the form into the matter. Bronze is a potential sphere, and the cloak is an actual sphere. But round bronze is equally the essence of both the actual sphere and the potential one. The bronze and the roundness are not two separate things. The bronze is potentially a sphere, and when it is made round it constitutes an actual one—a single sphere of bronze.

It is easy to see how this hylomorphic analysis explains the unity of a substantial material particular, since neither the matter nor the form of such a particular is by itself a single material individual, and it is only when they are taken together that they constitute such an individual. But the question Aristotle is trying to answer is this: "why on earth is something one when the account of it is what we call a defnition?". Since proper definables are universals, it remains to be seen how the proposed solution applies to them. After all, universals are not material objects, and so it is not clear how they can be viewed as hylomorphic compounds. But Aristotle has at his disposal a concept that can fill this bill perfectly, viz., the concept of intelligible matter. So we surmise that it is for this reason that Aristotle goes on to introduce matter into the current context. If this is so, we may conclude that the material component in the definition of a species is intelligible matter. Elsewhere, he explicitly describes genus as matter: "the genus is the matter of what it is said to be the genus of". So a species too, although it is not itself a material object, can be considered a hylomorphic compound. Its matter is its genus, which is only potentially the species defined; its differentia is the form that actualizes the matter. The genus does not actually exist independently of its species any more than bronze exists apart from all form. The genus animal, for example, is just that which is potentially some specific kind of animal or other. Aristotle concludes "the ultimate matter and the form are one and the same, the one potentially, the other actively ... and what potentially is and what actively is are in a way one."

This solution, of course, applies only to hylomorphic compounds. But that is all it needs to do, according to Aristotle. For he ends the chapter by claiming that the problem of unity does not arise for other kinds of compounds that are not material: "Things that have no matter ... are all unconditionally just what is a one".

Theology

The science of being qua being is a science of form. But it is also theology, the science of god. The question now is, how can it be both? And to it Aristotle gives a succinct answer:

> If there is some immovable substance, this will be prior and will be primary philosophy, and it will be universal in this way, namely, because it is primary. And it will belong to it to get a theoretical grasp on being qua being, both what it is and the things that belong to it insofar as it is being.

So the primacy of theology, which is based on the fact that it deals with substance that is eternal, immovable, and separable, is supposedly what justifies us in treating it as the universal science of being qua being.

A reminder, first, of what this primacy is. As we saw in Sections 2–3 above, only beings in the category of substance are separable, so that they alone enjoy a sort of ontological priority that is both existential and explanatory. Thus walking and being healthy are characterized as "incapable of being separated," on the grounds that there is some particular substantial underlying subject of which they are predicated. Often, indeed, separability is associated with being such a subject: "The underlying subject is prior, which is why the substance is prior"; "If we do not posit substances to be separated, and in the way in which particular things are said to be separated, we will do away with the sort of substance we wish to maintain". Similarly, not being separable is associated with being predicated of such a subject: "All other things are either said of primary substances as subjects or in them as subjects. Therefore, if there were no primary substances, there could not be anything else". The starting-points and causes of all beings, then, must be substances. But for all that has been shown so far, the universe could still be made up of lots of separate substances having little ontologically to do with each other.

Here it may serve to return to which opens by calling attention to something said (legomenon) about substance, namely that:

> Something is said to be (legetai) substance, if not in more ways, at any rate most of all in four. For the essence, the universal, and the genus seem to be the substance of each thing, and fourth of these, the underlying subject.

Since "the primary underlying subject seems most of all to be substance", because what is said or predicated of it depends on it, the investigation begins with this subject, quickly isolating three candidates: the matter, the compound of matter and form, and the form itself, which is identical to essence. Almost as quickly, the first two candidates are at least provisionally excluded. A—perhaps the—major ground for their exclusion is the primacy dilemma, which we shall now briefly investigate.

The philosophical background to the dilemma is this. If you are a realist about scientific knowledge and truth, as Aristotle is, the structure of your scientific theories must mirror the structure of reality, so that scientific starting-points or first principles, must also be the basic building blocks of reality. Suppose that this is not so. Suppose that your physics tells you that atoms are the basic building blocks of reality and that your psychology tells you that sense-perceptions are the starting-points of scientific knowledge. Then you will face a very severe problem, that of skepticism. For a wedge can be driven between the starting-points of scientific knowledge and reality's basic building blocks. René Descartes' famous dreaming argument is one familiar form such a wedge might take. Your sense-perceptions are consistent with your being always asleep and having a very detailed dream.

Aristotle introduces a similar problem about the relation between our scientific representation of the world and how the world in itself is structured:

> We must … ask whether [the starting-points] are universal or exist in the way we say particulars do. For if they are universal, they will not be substances. For no common thing signifies a this something but a such-and-such sort of thing, whereas substance is a this something.… If then the starting-points are universals, these things follow. But if they are not universals, but as particulars, they will not be scientifically knowable. For scientific knowledge of all things is universal. Thus there will be other starting-points prior to the starting-points, namely, those that are predicated universally, if indeed there is going to be scientific knowledge of these.

The basic building blocks of reality, (Aristotelian) science tells us, are particular matter-form compounds. Yet science's own starting-points are the forms—the universal essences—of such things. There is no science of you, or of me, though there is one of human beings. How, then, can science possibly be reflecting accurately the structure of reality, when its starting-points and those of reality fail so radically to map onto each other? For there is no greater difference, it seems, than that between particulars and universals. The thing to do, then, given that science provides our best access to the nature of reality, is to investigate the universal forms or essences that are basic to it.

Aristotle begins the investigation with the most familiar and widely recognized case, which is the form or essence present in sublunary matter-form compounds. It is announced, but not begun until some chapters later and not really completed until the end. And by then it is with actuality (entelecheia) or activity (energeia) that form is identified, and matter with potentiality. The science of being qua being can legitimately focus on form, or actuality, then, as the factor common to all substances, and so to all the beings. But unless it can be shown that there is some explanatory connection between the forms of all these substances, the non-fragmentary nature of being itself will still not have been established, and the pictures given to us by the various sciences will, so to speak, be separate pictures, and the being they collectively portray, divided.

The next stage in the unification of being, and the legitimation of the science dealing with it qua being, is effected by an argument in Λ.6 that trades on the identification of form with actuality and matter with potentiality:

If there is something that is capable of moving things or acting on them, but that is not actively doing so, there will not be movement, since it is possible for what has a capacity not to activate it. There is no benefit, therefore, in positing eternal substances, as those who accept the Forms do, unless there is to be present in them some starting-point that is capable of causing change. Moreover, even this is not enough, and neither is another substance beyond the Forms. For if it will not be active, there will not be movement. Further, even if it will be active, it is not enough, if the substance of it is a capacity. For then there will not be eternal movement, since what is potentially may possibly not be. There must, therefore, be such a starting-point, the very substance of which is activity. Further, accordingly, these substances must be without matter. For they must be eternal, if indeed anything else is eternal. Therefore they must be activity.

Matter-form compounds are, as such, capable of movement and change. The canonical examples of them—perhaps the only genuine or fully fledged ones—are living metabolizing beings. But if these beings are to be actual, there must be substances whose very essence is activity—substances that do not need to be activated by something else.

With matter-form compounds shown to be dependent on substantial activities for their actual being, a further element of vertical unification is introduced into beings, since layer-wise the two sorts of substances belong together. Laterally, though, disunity continues to threaten. For as yet nothing has been done to exclude the possibility of each compound substance having a distinct substantial activity as its own unique activator. Being, in that case, would be a set of ordered pairs, the first member of which would be a substantial activity, the second a matter-form compound, with all its dependent attributes.

In Metaphysics Λ.8 Aristotle initially takes a step in the direction of such a bipartite picture. He asks how many substantial activities are required to explain astronomical phenomena, such as the movements of the stars and planets, and answers that there must be forty-nine of them. But these forty-nine are coordinated with each other so as to form a system. And what enables them to do so, and to constitute a single heaven, is that there is a single prime mover of all of them:

It is evident that there is but one heaven. For if there are many, as there are many human beings, the starting-point for each will be one in form but in number many. But all things that are many in number have matter, for one and the same account applies to many, for example, human beings, whereas Socrates is one. But the primary essence does not have matter, since it is an actuality. The primary immovable mover, therefore, is one both in account and in number. And so, therefore, is what is moved always and continuously. Therefore, there is only one heaven.

What accounts for the unity of the heaven, then, is that the movements in it are traceable back to a single cause: the prime or primary mover.

Leaving aside the question of just how this primary mover moves what it moves directly, the next phase in the unification of beings is the one in which the sublunary world is integrated with the already unified superlunary one studied by astronomy. This takes place in Λ.10. One obvious indication of this unification is the dependence of the reproductive cycles of plants and animals on the seasons, and their dependence, in turn, on the movements of the sun and moon. And beyond even this there is the unity of the natural world itself, which is manifested in the ways in which its inhabitants are adapted to each other:

All things are jointly organized in a way, although not in the same way—even swimming creatures, flying creatures, and plants. And the organization is not such that one thing has no relation to another but rather there is a relation. For all things are jointly organized in relation to one thing—but it is as in a household, where the free men least of all do things at random, but all or most of the things they do are organized, while the slaves and beasts can do a little for the common thing, but mostly do things at random. For this is the sort of starting-point that the nature is of each of them. I mean, for example, that all must at least come to be disaggregated; and similarly there are other things which they all share for the whole.

Thus the sublunary realm is sufficiently integrated with the superlunary one that we can speak of them as jointly having a nature and a ruler, and as being analogous to an army and a household.

We may agree, then, that the divine substances in the superlunary realm and the compound substances in the sublunary one have prima facie been vertically integrated into a single explanatory system. As a result, when we look at the form of a sublunary matter-form compound, we will find in it the mark of a superlunary activator, just as we do in the case of the various heavenly bodies, and, as in the line of its efficient causes, we find "the sun and its movement in an inclined circle". Still awaiting integration, though, are mathematical objects, such as numbers. But in Books M and N these are shown to be not substantial starting-points and causes but abstractions from perceptible sublunary beings—they are dependent entities, in other words, rather than self-subsistent ones. Similarly, in Physics we read:

The mathematician too busies himself about these things, although not insofar as each of them is the limit of a natural body, nor does he get a theoretical grasp on the coincidents of natural bodies insofar as they are such. That is why he separates them. For they are separable in the understanding from movement, and so their being separated makes no difference, nor does any falsehood result from it.

This completes the vertical and horizontal unification of being: attributes depend on substances, substantial matter-form compounds depend on substantial forms, or activities, numbers depend on matter-form compounds.

Beings are not said to be "in accord with one thing," therefore, as they would be if they formed a single first-order genus, but "with reference to one thing," namely, a divine substance that is in essence an activity. And it is this more complex unity, compatible with generic diversity, and a genuine multiplicity of distinct first-order sciences, but just as robust and well grounded, that grounds and legitimates the science of being qua being as a single science dealing with a genuine object of study. The long argument that leads to this conclusion is thus a sort of proof of the existence, and so of the possibility, of the science on which the Metaphysics focuses. It is also the justification for the claim, which we looked at before, that the science of being qua being is in fact theology.

There, then, in the starry heavens above us, are the forty-nine celestial spheres, all moving eternally in fixed circular orbits. The outermost one, which contains all the others, is the primary heaven. Questions immediately arise: (i) how is the primary heaven moved by the primary mover, the primary god? Aristotle gives his response in Λ.7:

There is something that is always moved with an unceasing movement, which is in a circle (and this is clear not from argument alone but also from the facts). So the primary heaven would be eternal. There is, therefore, also something that moves it. But since what is moved and moves something is something medial, there is something that moves without being moved, being eternal, substance, and activity. This, though, is the way the object of desire and the intelligible object move things: they move them without being moved. Of these objects, the primary ones are the same.

Thus the primary heaven is moved by the primary god, in the way that we are moved by a good that we desire. But how can the primary god be such a good? Moreover, why is he not moved by something else again?

The answer to question is also found:

Active understanding, though, is intrinsically of what is intrinsically best, and the sort that is to the highest degree best is of what is to the highest degree best. The understanding actively understands itself by partaking of the intelligible object. (For it becomes an intelligible object by touching and understanding one, so that understanding and intelligible object are the same.) For what is receptive of the intelligible object and of the substance is the understanding, and it is active when it possesses it, so that this rather than that seems to be the divine thing that understanding possesses, and contemplation seems to be most pleasant and best. If, then, that good state, which we are sometimes in, the god is always in, that is a wonderful thing, and if to a higher degree, that is yet more wonderful. But that is his state. And life, too, certainly belongs to him. For the activity of understanding is life, and he is that activity; and his intrinsic activity is life that is best and eternal.

What the primary heaven is moved by, then, is the wish to be in the good state of active contemplation that we, when we are happiest, are in, and that the primary god is always in because he just is that activity. Just as we seek the good that the primary god is, so too does the primary heaven and its forty-eight celestial companions.

This brings us to question—When the understanding is actively contemplating something, that something—that intelligible object—is what activates it. So why isn't that object yet more primary than the primary god? Aristotle gives his answer in Λ.9; the reasoning, though compressed, should now be fairly readily intelligible:

What does it understand? For it is either itself or something else. And if something else, then either always the same thing or sometimes this and sometimes that. Does it, then, make a difference or none at all whether it actively understands the good or some random object? Or are there not certain things that it would be absurd for it to think of? It is clear, therefore, that it actively understands what is most divine and most estimable and does not change, since change would be for the worse, and would already be a sort of movement. First, then, if its substance is not active understanding but rather a capacity, … it is clear that something else would be more estimable than the understanding, namely, what is understood. And indeed to understand and active understanding will belong even to someone who actively understands the worst thing, so that if this is to be avoided, the active understanding would not be the best thing. It is itself, therefore, that it understands, if indeed it is the most excellent thing, and the active understanding is active understanding of active understanding.

God is the understanding that understands himself, because his understanding is like ours would be if we imagine it as being the intelligible equivalent of seeing light without seeing any other visible object. From the inside, then, from the point of view of the subject experiencing it, it is a state of consciousness of a sort familiar from the writings of the great religious mystics, in which both subject and object disappear from an awareness that yet remains fully and truly attentive, fully alive and joyous. Insofar as we have any experience-based evidence of what a beatific state is like, this one surely approximates to it. Were we to experience it or something like it, then, there is some reason to think that we would agree that it is bliss indeed, blessed happiness unalloyed. This is the conclusion Aristotle himself comes to and defends in Nicomachean Ethics X.6–8. Practical wisdom and theoretical wisdom, it follows, have the same ultimate starting-point, the same first principle, so that wisdom, too, is something unified.

Go back now to the primacy dilemma and notice that its resolution is within our grasp, though one might be forgiven for not readily understanding Aristotle's statement of it in M.10:

The fact that all scientific knowledge is universal, so that the starting-points of beings must also be universal and not separate substances, involves the greatest puzzle of those mentioned. But though there is surely a way in which what is said is true, there is another way in which it is not true. For scientific knowledge, like knowing scientifically, is twofold, one potential, the other active: the capacity, being as matter, universal and indefinite, is of what is universal and indefinite, whereas the activity, being definite, is of what is definite—being a this something of a this something. But it is only coincidentally that sight sees universal color, because this color that it sees is a color, and so what the grammarian theoretically grasps, namely, this A, is an A. For if the starting-points must be universal, what comes from them must also be universal, as in the case of demonstrations. And if this is so, there will be nothing separable and no substance either. However, in one way scientific knowledge is universal, but in another it is not.

The idea is this. Since forms or essences are universals, you and I may both know the same form, as we may both know the letter A. But when I actively know or contemplate that universal form, what is now before my mind is a particular: this actualization of that universal. Now consider the primary god. He is eternally and essentially the object of the active understanding that he is. So he is a substantial particular. But since he is essentially an activity, he is also a universal essence of a special sort—one that can only be actual, never merely potential. In a way, then, the primary god overcomes the difference between particulars and universals that seemed unbridgeable. For he is at once a concrete particular and the starting-point of all scientific knowledge. He thereby unifies not just being, but the scientific knowledge of it as well, insuring that the latter fits the former in the way that realism requires.

5

THOMISM

5.1 INTRODUCTION

The advent of Christianity had important effects in philosophy as in other aspects of human life. Initially, Christians were opposed to philosophical claims of any kind; they saw philosophy as an essentially pagan phenomenon and refused to allow the propriety of subjecting Christian dogma to philosophical scrutiny. Christian truth rested on revelation and did not need any certificate of authenticity from mere reason. Later, however, attempts were made to produce a specifically Christian metaphysics, to think out a view of the universe and of humans' place in it that did justice to the Christian revelation and nevertheless rested on arguments that might be expected to convince Christians and non-Christians alike. St. Thomas Aquinas was only one of a number of important thinkers in medieval times who produced Christian philosophies; others—such as the philosophers John Duns Scotus in the late 13th century and William of Ockham in the first half of the 14th century—took significantly different views. In selecting the system of Aquinas for summary here, the factor that has weighed most has been its influence, particularly in postmedieval times. Aquinas was not the only medieval philosopher of distinction, but Thomism is alive as other medieval systems are not.

The central claim of Thomism is that reflection on everyday things and the everyday world reveals it as pointing beyond itself to God as its sustaining cause. Ordinary existents, such as human beings, are in the process of constant change. The change, however, is not normally the result of their own efforts, and, even when it is, it does not depend on them exclusively. No object in the familiar world can fully account for its own esse (i.e., its own act of existing), nor is it wholly self-sufficient; all are affected from without, or at least operate in an environment that is not of their own making. To say this is to say that they are one and all finite. Although finite things can be, and commonly are, stimulated to activity or kept in activity by other finite things, it does not follow that there might be finite things and nothing else. On the contrary, the finite necessarily points beyond itself to the infinite. The system of limited beings, each dependent for its activity on something else of the same kind, demands for its completion the existence of an unlimited being, one that is the source of change in other things but is not subject to change itself. Such a being would be not a cause like any other but a first or ultimate cause; it would be the unconditioned condition of the existence of all other things. Aquinas believed that human reason can produce definitive proofs of the existence of an infinite or perfect being, and he had no hesitation in identifying that being with the Christian God. Because, however, the movement of his thought was from finite to infinite, he claimed to possess only so much philosophical knowledge of the Creator as could be arrived at from study of his creation. Positive knowledge of the divine nature was not available; apart from revelation, one could only say what God is not, or conceive of his attributes by the imperfect method of analogy.

Aquinas worked out his ideas at a time when the philosophy of Aristotle was again becoming familiar in western Europe after a period of being largely forgotten, and many of his detailed theories show Aristotelian influence. He assumed the general truth of the Aristotelian picture of the natural world and the general correctness of Aristotle's way of interpreting natural phenomena. He also took over many of Aristotle's ideas in the fields of ethics and politics. He gave the latter, however, a distinctively different twist by making the final end of human beings not philosophical contemplation or the exercise of virtue in the political sphere but the attainment of the beatific vision of God; it was Christian rather than Greek ideas that finally shaped his view of the "summum bonum". Similarly, his celebrated proofs of God's existence proceeded against a background that is obviously Aristotelian but that need not be presupposed for their central thought to have validity. Thomism can certainly be seen, and historically must be seen, as the system of Aristotle adapted to Christian purposes. It is important, however, to stress that the adaptation resulted in something new, a distinctive way of looking at the world that still has its adherents and still commands the respect of philosophers.

Thomism is the philosophical and theological school that arose as a legacy of the work and thought of Thomas Aquinas (1225–1274), the Dominican philosopher, theologian, and Doctor of the Church. In philosophy, Aquinas' disputed questions and commentaries on Aristotle are perhaps his best-known works.

In theology, his Summa Theologica is amongst the most influential documents in medieval theology and continues to be the central point of reference for the philosophy and theology of the Catholic Church. In the 1914 motu proprio Doctoris Angelici, Pope Pius X cautioned that the teachings of the Church cannot be understood without the basic philosophical underpinnings of Aquinas' major theses:

The capital theses in the philosophy of St. Thomas are not to be placed in the category of opinions capable of being debated one way or another, but are to be considered as the foundations upon which the whole science of natural and divine things is based; if such principles are once removed or in any way impaired, it must necessarily follow that students of the sacred sciences will ultimately fail to perceive so much as the meaning of the words in which the dogmas of divine revelation are proposed by the magistracy of the Church.

5.2 METAPHYSICS ACCORDING TO THOMAS AQUINAS

The name of metaphysics stands for a broad range of human activities and literature. Perhaps the only binding and distinguishing characteristic of this range is that of somehow relating to what, in principle, cannot be seen or felt or heard or in any way sensed. Metaphysics sometimes encompasses inquiry into angels and demons, purely rationalistic notions about God, or the exploration of occult forces of nature and of the human spirit. The great bulk of what passes for metaphysics in the popular mind bears little resemblance or relation to genuinely rigorous, scientific or philosophic reasoning. Indeed, even among some of the philosophically rigorous, anything bearing the name metaphysics receives an intellectually cool reception precisely because of the association with the non-material that is inextricably worked into its name.

The problem of defining the nature of metaphysics also may be seen by examining it in relation to other sciences. Given the fact that everything that one can know either is sensed or knowledge of it is derived from the senses, and so is in some way dependent on physical reality, there arises the problem of finding a subject- matter unique to metaphysics. Granted that all sciences study things, either some class of material things or some aspect of them (e.g. physics studies changing being; mathematics, quantified being; logic, beings of reason) what does metaphysics study that is not studied by physics, mathematics or logic, but is unique to itself? Or put another way, what does the metaphysician know about things that the physicist, or mathematician or logician does not?

The answer given by Aristotle, and by St. Thomas Aquinas after him, is that metaphysics studies being as being. This implies, and the diversity of the other sciences confirms, that being can be studied in some way other than as being, e.g. as changing, etc. But the fact remains that everything we directly experience through our external senses is changeable being, e.g. rocks and trees. And following Aristotle's analysis, things are changeable insofar as they are material. As essentially changeable, such objects of sensation are popularly described as being made up out of "stuff."

"The term (matter) pretty well corresponds in its use to some uses of the word 'stuff' in English. …When the stuff that Miss T. eats turns into Miss T., Aquinas would say that the same matter was first say bread and then part of a human body".

So, we may inquire how one can come to know being as distinct from stuff, if we take "stuff" to mean the matter of the things which we see and touch. And it is not immediately obvious that being means anything other than the things which are made out of some common stuff. So the question becomes: How can metaphysics study being as being in contradistinction from physics, which studies being as material and changing, if all we can directly know is changing being?

In the thought and writings of St. Thomas Aquinas, metaphysics holds an honored place among the speculative rational sciences and philosophical disciplines. It is the queen of rational disciplines and receives the name of wisdom. In the Prooemium to his Commentary on the Metaphysics of Aristotle, Aquinas explains why the science enjoys the preeminence that it does. He also endeavors to give a rationale for how the science ought to be named, since this field of inquiry, so far only referred to as metaphysics, traditionally has borne other names, i.e. First Philosophy, and Theology. Now, while this philosophical enterprise involves immateriality in all three of the aspects to which these names correspond, that aspect that is the science's core is a treatment of immateriality in a qualified sense.

According to Aquinas, three features characterize speculative, philosophical wisdom: being most causal/certain, most universal, and most immaterial. These features correspond to the realities that are treated by the science: God, being and its principles, and subsistent immaterial realities (i.e. angels and the rational soul). Likewise, to these features there also correspond three names. The most intelligent speculative science which orders all others is called First Philosophy insofar as it explains all things in terms of the first cause of things. Insofar as the most intelligent science seeks to explain being and its principles, philosophical wisdom carries the name of Metaphysics. And finally, theology is the name of the most intelligent science according as this speculative wisdom examines immaterial subsistents (e.g. God and angels) through their causal relation to material things.

But the question remains as to which name is most proper, and which feature most essential to this science. Now the investigations of First Philosophy and philosophical theology are conducted only by the relation they have to the subject of metaphysics. What characterizes First Philosophy and philosophical theology enters into the consideration of wisdom only because God and immaterial subsistents are related to being and its principles. The pursuer of speculative wisdom studies the First Principle and separate intellectual substances only in order to explain the most universal characteristics of all things. Thus, Aquinas concludes that speculative wisdom is essentially the study of the most universal aspect of real things, their being.

And so, Metaphysics is the proper name of this speculative philosophical wisdom, and not First Philosophy nor Theology although only these latter two were used by Aristotle himself to characterize wisdom, and not metaphysics. Thus, positively immaterial substances, i.e. God and intellectual substance, are not the proper subject of metaphysics. Rather being as being ens inquantum ens or being in general or common being ens commune is the subject matter of metaphysics. Aquinas concludes to ens commune as the subject of metaphysics because ens commune is the genus, i.e. the intelligible formality, which metaphysics hopes to explain by appealing to positively immaterial things.

Consequently, it must be the office of one and the same science to consider separate substances and being- in-general ens commune which is the genus of which the separate substances mentioned above are the common and universal causes.... For the subject of a science is the genus whose causes and properties we seek, and not the causes themselves of the particular genus studied, because the knowledge of the causes of some genus is the goal to which the investigation of the science attains.

Immaterial realities, things having in themselves no "stuff," become known within the science of metaphysics as the common and universal causes of being in general. They fall within the science only in order to explain ens commune. However, knowledge of the immaterial is necessary in order for ens commune to be known. That is, the nature of being as being requires knowledge of some immaterial realities outside and prior to the science of metaphysics, though this point is not explicitly asserted in the prooem. to Aquinas' Commentary on the Metaphysics.

Aquinas thus applies the name metaphysics to the science of speculative wisdom in virtue of the subject matter of the science, being in general. Yet he claims that being in general is predicated of those things separate from matter, i.e. stuff, both in intelligible constitution ratio and in being esse.

Now although the subject of the science is being in general, the whole of it is predicated of those things which are separate from matter both in their intelligible constitution and in being esse. For it is not only those things which can never exist in matter that are said to be separable from matter in their intelligible constitution and in being, such as God and the intellectual substances, but also those things that can exist without matter, such as being in general. This could not be the case, however, if their being depended on matter.... It is called metaphysics inasmuch as it considers being and the attributes which naturally accompany being (for things which transcend the physical order are discovered by the process of analysis, as the more common is discovered after the less common).

5.2.1 Ibid.

Ens commune is immaterial to the extent that it names that aspect of things that is not strictly identifiable with materiality, although every material thing may be considered as having this aspect. Ens commune is immaterial insofar as within its designation fall all things, material and immaterial, yet it is not the consideration of these things in their materiality or immateriality respectively. As Aquinas puts it in his commentary on Boethius' De Trinitate, something can exist separate from matter and motion ... because by its nature it does not exist in matter and motion; but it can exist without them, though we sometimes find it with them".

And that which is separate from matter in this sense is the subject matter of metaphysics.

So, while the subject matter of metaphysics, ens commune, is immaterial, it is a negative immateriality, i.e. it is the immateriality of considering all things, even material things, apart from any consideration of materiality as such. God and the angels can only be considered apart from matter since they are positively immaterial, but material things can be viewed both with regard to their material constitution, i.e. insofar as they move and change and are made of stuff, and with regard to their being, or unity, or act.

What is essential to metaphysics is its subject matter. As Aquinas expounds on this subject matter, ens commune, it is negatively immaterial. Ens commune as the subject of metaphysics must be negatively immaterial for two reasons. First, being qua being is a formality under which one may consider things that is different from the formality of materiality. Thus, in order to do metaphysics, one must gain a perspective on all things, even material things, that is itself not a material perspective; one must view the world with an eye toward being. The second reason that the subject of metaphysics must be negatively immaterial is that metaphysics is a science distinct from physics. The metaphysician must know being as distinct from materiality because this science which he is endeavoring to undertake, metaphysics, is distinct from that science, natural philosophy, which is the study of things as material.

Metaphysics is the study of the basic structure of reality. It is, in Aristotle's words, the study of being as being, rather than the study of any particular being per se. Metaphysics is the framework by which we understand reality. We can't avoid metaphysics — every act of understanding entails a metaphysical framework, a perspective. One might say that our metaphysical perspective is that by which we understand, contrasted to nature itself, which is that which we understand.

Our own metaphysical framework is often opaque to us. We use it, like we might use an intuitive political bias, without really examining the framework we are using. We each have a metaphysical bias— it's unavoidable, and the important question is: does our bias lead us toward or away from the truth? Gaining metaphysical insight is not easy, but it pays big dividends. It helps us to know the truth— indeed, it is that by which we know the truth.

5.3 A RIGOROUS AND CONSISTENT SYSTEM

St. Thomas Aquinas developed a rigorous and consistent system of metaphysics. He was the first Christian philosopher to insist that faith and reason, properly understood, are never in conflict. Belief in God is not contrary to knowledge of the natural world. St. Thomas' doctrine was controversial in his day, but it was accepted by the Church in the centuries after his death, and it became the intellectual foundation of the modern world, including the cornerstone of modern science.

Ironically, the correspondence of faith and reason is controversial today, especially in the atheist community. The denial of the compatibility between faith and reason is a lynchpin of atheist arguments for naturalism: atheists insist that science tells us the real truth about the world, and faith in God is superstition. The Thomistic reply is that genuine faith and reason both point to the same truth. The Thomistic understanding of reason and its correspondence with faith offer a powerful reply to atheistic naturalism. For readers who are interested in metaphysics and in these modernist controversies, it's worthwhile taking a closer look at the principle that is the cornerstone of Thomistic metaphysics.

5.3.1 The Nature of Metaphysics

Saint Thomas, that is, Aquinas, clarifies the nature of metaphysics through ascertaining its particular subject-matter, its field of investigation. In order to ascertain the subject-matter of any particular science, Thomas distinguishes between the different intellectual operations that we use when engaged in some particular scientific endeavor. Broadly speaking, these fall into two categories: the speculative and the practical. Concerning some sciences, the intellect is merely speculative by contemplating the truth of some particular subject-matter; while concerning other sciences, the intellect is practical, by ascertaining the truth and seeking to apply. There are thus correspondingly two distinct classes of science: speculative science and practical science. Speculative sciences are those that contemplate truth whereas practical sciences are those that apply truth for some practical purpose. The sciences are then further distinguished through differentiating their various subject-matters.

Insofar as the speculative sciences merely contemplate truth but do not apply it for some practical purpose, the subject-matter of the speculative sciences is that which can be understood to some extent. Working within the Aristotelian tradition, Thomas holds that something is understood when it is separated from matter and is necessary to thing in some respect. For instance, when we understand the nature of a tree, what we understand is not primarily the matter that goes to constitute the tree in question, but what it is to be a tree, or the structuring principle of the matter that so organizes it and specifies it as a tree rather than a plant. Furthermore, assuming our understanding is correct, when we understand a thing to be a tree, we do not understand it to be a dog, or a horse, or a cat. Thus, in our understanding of a tree, we understand that which is necessary for the tree to be a tree, and not of anything that is not a tree. Hence, our understanding of a thing is separated from its matter and is necessary to it in some respect. Now, what is in motion is not necessary, since what is in motion can change. Thus, the degree to which we have understood something is conditional upon the degree to which it is separated from matter and motion. It follows then that speculative objects, the subject-matter of the speculative-sciences, insofar as they are what are understood, will be separated from matter and motion to some degree. Any distinctions that obtain amongst speculative objects will in turn signify distinctions amongst the sciences that consider those objects; and we can find distinctions amongst speculative objects based upon their disposition towards matter and motion.

There are three divisions that can apply to speculative objects, thereby permitting us to differentiate the sciences that consider such objects: (i) there is a class of speculative objects that are dependent on matter and motion both for their being and for their being understood, for instance, human beings cannot be without matter, and they cannot be understood without their constituent matter; (ii) there is a class of speculative objects that depend on matter and motion for their being, but not for their being understood, for instance, we can understand lines, numbers, and points without thereby understanding the matter in which they are found, yet such things cannot be without matter; (iii) there is a class of speculative objects that depend on matter and motion neither for their being nor for their being understood.

Given these three classes of speculative objects, the speculative sciences that consider them can be enumerated accordingly: (i) physical science considers those things that depend on matter and motion both for their being and for their being understood; (ii) mathematics considers those things that depend on matter and motion for their being but not for their being understood; (iii) metaphysics or theology deals with those things that depend on matter and motion neither for their being nor for their being understood. Before going on to consider the subject-matter of metaphysics in a little more detail, it is important to point out that Thomas takes this division of the speculative sciences as exhaustive. For Thomas, there could be no fourth speculative science; the reason for this is that the subject-matter of such a science would have to be those things that depend on matter and motion for their being understood but not for their being, for all other combinations have been exhausted. Now, if a thing depends on matter and motion for its being understood but not for its being, then matter and motion would be put into its definition, which defines a thing as it exists. But if a thing's existence is so defined as to include matter and motion, then it follows that it depends on matter and motion for its being; for it cannot be understood to be without matter and motion. Hence, all things that include matter and motion in their definitions are dependent on matter and motion for their being, but not all things that depend on matter and motion for their being depend on matter and motion for their being understood. There could be no fourth speculative science since there is no fourth class of speculative objects depending on matter and motion for their being understood but not for their being. Thomas thus sees this threefold division of the speculative sciences as an exhaustive one.

The third class of speculative objects comprises the objects of metaphysics or theology. Now Thomas does not equate these two disciplines, but goes on to distinguish between the proper subject-matter of metaphysics and the proper subject-matter of theology. Recall that this third class of speculative objects comprises those things depending on matter and motion neither for their being nor for their being understood. Such things are thus immaterial things; however, Thomas here draws a distinction. There are things that are immaterial insofar as they are in themselves complete immaterial substances; God and the angels would be examples of such things. To give the latter a title, let them be called positively immaterial. On the other hand there are things that are immaterial insofar as they simply do not depend on matter and motion, but can nevertheless be sometimes said to be found therein. In other words, things of the latter category are neutral with respect to being found in matter and motion, and hence they are neutrally immaterial. St Thomas's examples of the latter are: being, substance, potency, form, act, one and many; such things can apply equally to material things and, to some extent, to positively immaterial things. Thus, the neutrally immaterial seem to signify certain aspects or modes of being that can apply equally to material and to immaterial things. The question then arises: what is the proper subject-matter of metaphysics: the positively immaterial or the neutrally immaterial?

According to Thomas, unaided human reason cannot have direct knowledge of the positively immaterial; this is because such things outstrip the human intellect's capacity to know. Nevertheless, direct knowledge of the positively immaterial is possible, but this will not be on the basis of unaided human reason; it will require that the positively immaterial reveal themselves to us in some way, in which case direct knowledge of the positively immaterial will be dependent on some sort of revelation. As it is a purely rational science, not dependent on or presupposing the truths of revelation, metaphysics will be a study of the neutrally immaterial aspects of things, that is, a study of those modes of being that apply to all beings, whether they are material or immaterial. Such a study will be in accord with the Aristotelian conception of metaphysics as a study of being qua being, insofar as the neutrally immaterial apply to all beings and are not restricted to a certain class of beings. However, Thomas does not adopt the Aristotelian phrase (being qua being) as the subject-matter of metaphysics, he offers his own term. According to Thomas, ens commune is the proper subject-matter of metaphysics. Through an investigation of ens commune, an investigation into the aspects of being common to all beings, the metaphysician may indeed come to a knowledge of the causes of being and might thereby be led to the affirmation of divine being, but this is only at the end of the metaphysical inquiry, not at the beginning. Thus, metaphysics for Aquinas is a study of ens commune where this is understood as the common aspects of being without which a thing could not be; it does not presuppose the existence of divine being, and may not even be led to an affirmation of divine being though Thomas of course offers several highly complex metaphysical arguments for the existence of divine being, but this should not be taken to be essential to the starting point of Thomistic metaphysics.

Metaphysics then is a study of the certain aspects common to all beings; and it is the task of the metaphysician to uncover the aspects of being that are indeed common and without which a thing could not be. There are certain aspects of being that are common insofar as they are generally applicable to all beings, and these are essence and existence; all beings exist and have an essence, hence metaphysics will be primarily concerned with the nature of essence and existence and their relationship to each other. Having completed an investigation into essence and existence, the metaphysician must investigate the aspects of being that are common to particular instances of being; and this will be a study of (i) the composition of substance and accident, and (ii) the composition of matter and form. The format of Thomistic metaphysics then takes a somewhat dyadic structure of descending generality: (i) essence and existence, (ii) substance and accident, (iii) matter and form. The format of the remainder of this article will be an investigation into these dyadic structures.

5.4 ESSENCE AND EXISTENCE

The cornerstone of Thomistic metaphysics is the doctrine of essence and existence. It is this: essence is absolutely distinct from existence. This doctrine, which St. Thomas was the first philosopher to assert unequivocally and demonstrate with rigor, has profound implications for our understanding of reality, of nature, of science and of God. What does St. Thomas mean in saying "essence is absolutely distinct from existence"?

First, definitions. Essence is that which makes something the sort of thing it is. It is, succinctly, all the characteristics that are knowable about something. The essence of a cat is everything about the cat that makes it a cat. Its cat-shape, it's furriness, its meow, its animality, etc. Some things about the cat—the fact, for instance, that at (unfortunate) times it can be a projectile launcher or a meal—are not parts of the essence of a cat. They are extraneous to it, although in rare circumstances, they may be true of it. You can see here where that modern notion of "essence" comes from. Essence is what's important about something, what tells us what something really is.

A general notion of essence is the following: essence is the definable nature of the thing that exists. Quite generally then, the essence of a thing is signified by its definition. The immediate question then is how the essence of a thing relates to its existence. In finite entities, essence is that which has existence, but it is not existence; this is a crude articulation of Thomas's most fundamental metaphysical teaching: that essence and existence are distinct in finite entities. A consideration then of essence and existence in Thomas's metaphysical thought will thus be a consideration of his fundamental teaching that essence and existence are distinct.

The most famous, and to a certain degree the most controversial, instance wherein Thomas argues for a distinction between a thing's essence and its existence. The context is a discussion of immaterial substances and whether or not they are composed of matter. In that passage, Thomas is concerned with a popular medieval discussion known as universal hylemorphism. St Bonaventure,

Thomas's contemporary, had held that insofar as creatures are in potency in some respect, they must be material in some respect, since on the Aristotelian account, matter is the principle of potency in a thing. Thus, creatures, even immaterial creatures, must be material in some respect, even if this materiality is nothing like our corporeal materiality.

Thomas takes the notion of universal hylemorphism to be absurd. Not only does it conflict with the common sayings of the philosophers, but also it is precisely as separated from matter and all material conditions that we deem separate (immaterial) substances separate, in which case they cannot be composed of matter. But if such substances cannot be composed of matter, what accounts for their potentiality? Such substances are not God, they are not pure act, they are in potentiality in some respect. So, if they are not material, then how are they in potency? Thomas is thus led to hold that they have an element of potentiality, but this is not the potency supplied by matter; rather, immaterial substances are composed of essence and existence, and it is the essence of the thing, standing in potency to a distinct act of existence, that accounts for the potentiality of creatures and thereby distinguishes them from God, who is not so composed. Thomas's argumentation for the distinction between essence and existence unfolds on three stages and for each stage there has been at least one commentator who has held that Thomas both intended and established the real distinction therein.

In the first stage Thomas argues as follows. Whatever does not enter into the understanding of any essence is composed with that essence from without; for we cannot understand an essence without understanding the parts of that essence. But we can understand the essence of something without knowing anything about its existence; for instance, one can understand the essence of a man or a phoenix without thereby understanding the existence of either. Hence, essence and existence are distinct.

This little paragraph has generated considerable controversy, insofar as it is unclear what sort of distinction Thomas intends to establish at this stage. Is it merely a logical distinction whereby it is one thing to understand the essence of a thing and another to understand its existence? On this account, essence and existence could well be identical in the thing yet distinct in our understanding thereof (just as 'Morning star' and 'Evening star' are distinct in our conceptual expressions of the planet Venus, yet both are identical with Venus). On the other hand, does Thomas attempt to establish a real distinction whereby essence and existence are not only distinct in our understanding, but also in the thing itself? Commentators who hold that this stage only establishes a logical distinction focus on the fact that Thomas is here concerned only with our understanding of essence and not with actual (real) things; such commentators include Joseph Owens and John Wippel. Commentators who hold that this stage establishes a real distinction focus on the distinction between the act of understanding a thing's essence and the act of knowing its existence, and they argue that a distinction in cognitional acts points to a distinction in reality; such commentators include Walter Patt, Anthony Kenny, and Steven Long.

In the second stage of argumentation, Thomas claims that if there were a being whose essence is its existence, there could only be one such being, in all else essence and existence would differ. This is clear when we consider how things can be multiplied. A thing can be multiplied in one of three ways: (i) as a genus is multiplied into its species through the addition of some difference, for instance the genus 'animal' is multiplied into the species 'human' through the addition of 'rational'; (ii) as a species is multiplied into its individuals through being composed with matter, for instance the species 'human' is multiplied into various humans through being received in diverse clumps of matter; (iii) as a thing is absolute and shared in by many particular things, for instance if there were some absolute fire from which all other fires were derived. Thomas claims that a being whose essence is its existence could not be multiplied in either of the first two ways (he does not consider the third way, presumably because in that case the thing that is received or participated in is not itself multiplied; the individuals are multiplied and they simply share in some single absolute reality). A being whose essence is its existence could not be multiplied (i) through the addition of some difference, for then its essence would not be its existence but its existence plus some difference, nor could it be multiplied (ii) through being received in matter, for then it would not be subsistent, but it must be subsistent if it exists in virtue of what it is. Overall then, if there were a being whose essence is its existence, it would be unique, there could only be one such being, in all else essence and existence are distinct.

Notice that Thomas has once again concluded that essence and existence are distinct. John Wippel takes this to be the decisive stage in establishing that essence and existence are really distinct. He argues that insofar as it is impossible for there to be more than one being whose essence is its existence, there could not be in reality many such beings, in which case if we grant that there are multiple beings in reality, such beings are composed of essence and existence. On the other hand, Joseph Owens has charged Wippel with an ontological move and claims that Wippel is arguing from some positive conceptual content, to the actuality of that content in reality. Owens argues that we cannot establish the real distinction until we have established that there is something whose essence is its existence. Given the existence of a being whose essence is its existence, we can contrast its existence with the existence of finite things, and conclude that the latter are composites of essence and existence; and so Owens sees the real distinction as established at stage three of Thomas's argumentation: the proof that there actually is a being whose essence is its existence.

Thomas begins stage three with the premise that whatever belongs to a thing belongs to it either through its intrinsic principles, its essence, or from some extrinsic principle. A thing cannot be the cause of its own existence, for then it would have to precede itself in existence, which is absurd. Everything then whose essence is distinct from its existence must be caused to be by another. Now, what is caused to be by another is led back to what exists in itself. There must be a cause then for the existence of things, and this because it is pure existence; otherwise an infinite regress of causes would ensue.

It is here that Owens believes that Thomas establishes the real distinction; since Thomas establishes that there exists a being whose essence is its existence. Consequently, we can contrast the existence of such a being with the existence of finite entities and observe that in the latter existence is received as from an efficient cause whereas in the former it is not. Thus, essence and existence are really distinct. However, it is important to note that on this interpretation, the real distinction could not enter into the argument for the existence of a being whose essence is its existence; for, on Owens's account, such argumentation is taken to establish the real distinction. If it can be shown then that Thomas's argumentation for the existence of a being whose essence is its existence does presuppose the real distinction, then Owens's views as to the stage at which the real distinction is established would be considerably undermined.

Having established that essence and existence are distinct and that there exists a being whose essence is its existence, Thomas goes on to conclude that in immaterial substances, essence is related to existence as potency to act. The latter follows insofar as what receives existence stands in potency to the existence that it receives. But all things receive existence from the being whose essence is its existence, in which case the existence that any one finite thing possesses is an act of existence that actuates a corresponding potency: the essence. Thomas has thus shown that immaterial substances do indeed have an element of potency, but this need not be a material potency.

Notice that here Thomas correlates essence and existence as potency and act only after he has concluded to the existence of a being whose essence is its existence (God). One wonders then whether or not essence and existence can be related as potency and act only on the presupposition of the existence of God. Regardless of his preferred method. Thomas could very well have focussed on the efficiently caused character of existence in finite entities (as he does in the opening lines of the argument for the existence of God), and argued that insofar as existence is efficiently caused (whether or not this is from God), existence stands to that in which it inheres as act to potency, in which case the essence that possesses existence stands in potency to that act of existence. Therefore, Thomas need not presuppose the existence of God in order to hold that essence and existence are related as potency and act; all he need presuppose is (i) that essence and existence are distinct and (ii) that existence is efficiently caused in the essence/existence composite.

5.4.1 Now for Existence

Existence is that a thing is, rather than what a thing is. The existence of a thing is different from the essence of a thing. I can know the essence of a rock, but it is the rock's existence by which I stub my toe. I can't stub my toe on essence, no matter how hard it is.

Prior to St. Thomas, many philosophers considered existence to be a property of something, part of its essence. We might say that my cat Fluffy's essence is that she is shaped like a cat, purrs and meows, likes to play with yarn, and exists.

5.4.2 An Utter Metaphysical Distinction

St. Thomas emphatically pointed out that existence is not, and cannot be, any part of essence. Existence and essence are metaphysically utterly distinct—existence is not a genus, in scholastic terms, but is above every genus. Existence is not a characteristic or property of a thing. It is something much more fundamental.

To understand what Aquinas is getting at, consider again my cat Fluffy. I will describe her to you: she is calico, weighs nine pounds, hates baths, purrs, says "meow" several times a day, is three years old, and is expecting kittens. If you want to know more about her, just ask. I can describe her in any degree of detail you would like.

Now tell me this: does she exist? I have given you her essence, to any level of detail you want, but the fact or fallacy of her existence is not knowable from knowing her essence. In fact, I don't have a cat. I have a dog. But I can describe my cat completely, and you still can't know if she really exists.

5.5 PARTICIPATION

Essence/existence composites merely have existence; whatever an essence/existence composite is, it is not its existence. Insofar as essence/existence composites merely have, but are not, existence, they participate in existence in order to exist. This is a second of Thomas's fundamental metaphysical teachings: whatever does not essentially exist, merely participates in existence. Insofar as no essence/existence composite essentially exists, all essence/existence composites merely participate in existence. More specifically, the act of existence that each and every essence/existence composite possesses is participated in by the essence that exists.

As a definition of participation, Thomas claims that to participate is to take a part (in) (partem capere) something. Following this definition, Thomas goes on to explain how one thing can be said to take a part in and thereby participate in another; this can happen in three ways.

Firstly, when something receives in a particular fashion what pertains universally to another, it is said to participate in that other; for example, a species ('man') is said to participate in its genus ('animal') and an individual (Socrates) is said to participate in its species ('man') because they (the species and the individual) do not possess the intelligible structure of that in which they participate according to its full universality.

Secondly, a subject is said to participate in the accidents that it has (for instance, a man is a certain colour, and thereby participates in the colour of which he is), and matter is said to participate in the formal structure that it has (for instance, the matter of a statue participates in the shape of that statue in order to be the statue in question).

Thirdly, an effect can be said to participate in its cause, especially when the effect is not equal to the power of that cause. The effect particularises and determines the scope of the cause; for the effect acts as the determinate recipient of the power of the cause. The effect receives from its cause only that which is necessary for the production of the effect. It is in this way that a cause is participated in by its effect.

In all of the foregoing modes of participation, to participate is to limit that which is participated in some respect. This follows from the original etymological definition of participation, that to participate is to take a part (in); for if to participate is merely to take a part in something, the participant will not possess the nature of the thing in which it participates in any total fashion, but only in partial fashion. What then can we conclude about the participation framework that governs essence and existence?

Essences exist, but they do not exist essentially, they participate in their acts of existence. Insofar as an essence participates in its act of existence, the essence limits that act of existence to the nature of the essence whose act it is; for the essence merely has existence, it is not existence, in which case its possession of existence will be in accord with the nature of the essence. The act of existence is thus limited and thereby individuated to the essence whose act it is. As a concrete application of this, consider the following. George Bush's existence is not Tony Blair's existence; when George Bush came into existence, Tony Blair did not come into existence, and when George Bush ceases to exist, Tony Blair will, in all likelihood, not cease to exist. George Bush's existence is not indexed to the existence of Tony Blair, in which case the existence of either George Bush or Tony Blair is not identical to the other. The act of existence then is individuated to the essence whose act it is, and this because the essence merely participates in, and thereby limits, the act of existence that it possesses.

5.5.1 Substance and Accident

The next fundamental metaphysical category is that of substance. According to Aquinas, substances are what are primarily said to exist, and so substances are what have existence but yet are not identical with existence. Aquinas's ontology then is comprised primarily of substances, and all change is either a change of one substance into another substance, or a modification of an already existing substance. Given that essence is that which is said to possess existence, but is not identical to existence, substances are essence/existence composites; their existence is not guaranteed by what they are. They simply have existence as limited by their essence.

Let us begin with a logical definition of substance, as this will give us an indication of its metaphysical nature. Logically speaking, a substance is what is predicated neither of nor in anything else. This captures the fundamental notion that substances are basic, and everything else is predicated either of or in them. Now, if we transpose this logical definition of substance to the realm of metaphysics, where existence is taken into consideration, we can say that a substance is that whose nature it is to exist not in some subject or as a part of anything else, but what exists in itself. Thus, a substance is a properly basic entity, existing per se (though of course depending on an external cause for its existence), and the paradigm instances of which are the medium sized objects that we see around us: horses, cats, trees and humans.

On the other hand there are accidents. Accidents are what accrue to substances and modify substances in some way. Logically speaking, accidents are predicated of or in some substance; metaphysically speaking, accidents cannot exist in themselves but only as part of some substance. As their name suggests, accidents are incidental to the thing, and they can come and go without the thing losing its identity; whereas a thing cannot cease to be the substance that it is without losing its identity.

Accidents only exist as part of some substance. It follows then that we cannot have un-exemplified properties as if they were substances in themselves. Properties are always exemplified by some substance, whereas substance itself is un-exemplifiable. For example, brown is always predicated of something, we say that x is brown, in which case brown is an accident. However, brown is never found to be in itself, it is always exemplified by something of which it is said.

Within Aquinas's metaphysical framework, substances can be both material (cats, dogs, humans) and immaterial (angels), but as noted above, the paradigm instances of substances are material substances, and the latter are composites of matter and form; a material substance is neither its matter alone nor its form alone, since matter and form are always said to be of some individual and never in themselves. It follows then that material substances have parts, and the immediate question arises as to whether or not the parts of substances are themselves substances. In order to address this issue, we must ask two questions: (i) while they are parts of a substance, are such parts themselves substances? and (ii) are the parts of a substance themselves things that can exist without the substance of which they are parts?

Concerning (i) we must say that whilst they are parts of a substance such parts cannot be substances; this is so given the definition of substance outlined above: that whose nature it is to exist not in some subject. Given that the parts of a substance are in fact parts of a substance, it is their nature to exist in some subject of which they are a part. Consequently, the parts of a substance cannot themselves be substances.

Concerning (ii) the case is somewhat different, now we must consider whether or not the parts of a substance can exist without the substance of which they are parts, that is, after the dissolution of the substance of which they are parts do the parts become substances in themselves? The parts of a substance receive their identity through being the parts of the substance whose parts they are. Thus, the flesh and bone of a human are flesh and bone precisely because they are parts of a human. When the human dies, the flesh and bone are no longer flesh and bone (except equivocally speaking) because they are no longer parts of a human substance; rather, the flesh and bone cease to function as flesh and bone and begin to decompose, in which case they are not themselves substances. However, on Aquinas's view, the elements out of which a substance is made can indeed subsist beyond the dissolution of the substance. Thus, whilst the elements are parts of the substance, they are not, as parts of a substance, substances in themselves, but when the substance dissolves, the elements will remain as independent substances in their own right. Thus, in the case of the dissolution of the human being, whilst the flesh and bone no longer remain but decompose, the elements that played a role in the formation of the substance remain. In more contemporary terms we could say that before they go to make up the bodily substances we see in the world, atoms are substances in themselves, but when united in a certain form they go to make cats, dogs, humans, and cease to be independent substances in themselves. When the cat or dog or human perishes, its flesh and bones perish with it, but its atoms regain their substantial nature and they remain as substances in themselves. So, a substance can have its parts, and for as long as those parts are parts of a substance, those parts are not substances in themselves, but when the substance decomposes, those parts can be considered as substances in themselves so long as they are capable of subsisting in themselves.

5.5.2 Matter and Form

A very crude definition of matter would be that it is the 'stuff' out of which a thing is made, whereas form is signified by the organisation that the matter takes. A common example used by Aquinas and his contemporaries for explaining matter and form was that of a statue. Consider a marble statue. The marble is the matter of the statue whereas the shape signifies the form of the statue. The marble is the 'stuff' out of which the statue is made whereas the shape signifies the form that the artist decided to give to the statue. On a more metaphysical level, form is the principle whereby the matter has the particular structure that it has, and matter is simply that which stands to be structured in a certain way. It follows from this initial account that matter is a principle of potency in a thing; since if the matter is that which stands to be structured in a certain way, matter can be potentially an indefinite number of forms. Form on the other hand is not potentially one thing or another; form as form is the kind of thing that it is and no other.

On Aquinas's account, there are certain levels of matter/form composition. On one level we can think of the matter of a statue as being the marble whereas we can think of the shape of the statue as signifying the form. But on a different level with can think of the marble as signifying the form and something more fundamental being the matter. For instance, before the marble was formed into the statue by the sculptor, it was a block of marble, already with a certain form that made it 'marble'. At this level, the marble cannot be the matter of the thing, since its being marble and not, say, granite, is its form. Thus, there is a more fundamental level of materiality that admits of being formed in such a way that the end product is marble or granite, and at a higher level, this formed matter stands as matter for the artist when constructing the statue.

If we think of matter as without any form, we come to the notion of prime matter, and this is a type of matter that is totally unformed, pure materiality itself. Prime matter is the ultimate subject of form, and in itself indefinable; we can only understand prime matter through thinking of matter as wholly devoid of form. As wholly devoid of form prime matter is neither a substance nor any of the other categories of being; prime matter, as pure potency, cannot in fact express any concrete mode of being, since as pure potency is does not exist except as potency. Thus, prime matter is not a thing actually existing, since it has no principle of act rendering it actually existing.

Matter can be considered in two senses: (i) as designated and (ii) as undesignated. Designated matter is the type of matter to which one can point and of which one can make use. It is the matter that we see around us. Undesignated matter is a type of matter that we simply consider through the use of our reason; it is the abstracted notion of matter. For instance, the actual flesh and bones that make up an individual man are instances of designated matter, whereas the notions of 'flesh' and 'bones' are abstracted notions of certain types of matter and these are taken to enter into the definition of 'man' as such. Designated matter is what individuates some form. As noted, the form of a thing is the principle of its material organisation. A thing's form then can apply to many different things insofar as those things are all organised in the same way. The form then can be said to be universal, since it remains the same but is predicated over different things. As signifying the actual matter that is organised in the thing, designated matter individuates the form to 'this' or 'that' particular thing, thereby ensuring individuals of the same form.

Given that form is the principle of organisation of a thing's matter, or the thing's intelligible nature, form can be of two kinds. On the one hand, form can be substantial, organising the matter into the kind of thing that the substance is. On the other hand, form can be accidental, organising some part of an already constituted substance. We can come to a greater understanding of substantial and accidental form if we consider their relation to matter. Substantial form always informs prime matter and in doing so it brings a new substance into existence; accidental form simply informs an already existing substance (an already existing composite of substantial form and prime matter), and in doing so it simply modifies some substance. Given that substantial form always informs prime matter, there can be only one substantial form of a thing; for if substantial form informs prime matter, any other form that may accrue to a thing is posterior to it and simply informs an already constituted substance, which is the role of accidental form. Thus, there can only be one substantial form of a thing.

As stated above, essence is signified by the definition of a thing; essence is the definable nature of the thing that exists. A thing's essence then is its definition. It follows that on Thomas's account the essence of a thing is the composition of its matter and form, where matter here is taken as undesignated matter. Contrary to contemporary theories of essence, Aquinas does not, strictly speaking, take essence to be what is essential to the thing in question, where the latter is determined by a thing's possessing some property or set of properties in all possible worlds. In the latter context, the essence of a thing comprises its essential properties, properties that are true of it in all possible worlds; but this is surely not Aquinas's view. For Aquinas, the essence of a thing is not the conglomeration of those properties that it would possess in all possible worlds, but the composition of matter and form. On a possible-worlds view of essence, the essence of a thing could not signify the matter/form composite as it is in this actual world, since such a composite could be different in some possible world and therefore not uniform across all possible worlds. Thus, Aquinas does not adopt a possible-worlds view of essence; he envisages the essence of a thing as the definition or quiddity of the thing existing in this world, not as it would exist in all possible worlds.

5.6 METAPHYSICS

Aquinas says that the fundamental axioms of ontology are the principle of non-contradiction and the principle of causality. Therefore, any being that does not contradict these two laws could theoretically exist, even if said being were incorporeal.

5.6.1 Predication

Aquinas noted three forms of descriptive language when predicating: univocal, analogical, and equivocal.

- Univocality is the use of a descriptor in the same sense when applied to two objects or groups of objects. For instance, when the word "milk" is applied both to milk produced by cows and by any other female mammal.
- Analogy occurs when a descriptor changes some but not all of its meaning. For example, the word "healthy" is analogical in that it applies both to a person or animal which enjoys good health and to some food or drink which promotes health.
- Equivocation is the complete change in meaning of the descriptor and is an informal fallacy, for example when the word "bank" is applied to river banks and financial banks. Modern philosophers call it ambiguity.

Further, the usage of "definition" that Aquinas gives is the genus of the being, plus a difference that sets it apart from the genus itself. For instance, the Aristotelian definition of "man" is "rational animal"; its genus being animal, and what sets apart man from other animals is his rationality.

5.6.2 Causality

Aristotle categorized causality into four subsets in the Metaphysics, which is an integral part of Thomism:

"In one sense the term cause means (a) that from which, as something intrinsic, a thing comes to be, as the bronze of a statue and the silver of a goblet, and the genera of these. In another sense it means (b) the form and pattern of a thing, i.e., the intelligible expression of the quiddity and its genera (for example, the ratio of 2:1 and number in general are the cause of an octave chord) and the parts which are included in the intelligible expression. Again, (c) that from which the first beginning of change or of rest comes is a cause; for example, an adviser is a cause, and a father is the cause of a child, and in general a maker is a cause of the thing made, and a changer a cause of the thing changed. Further, a thing is a cause (d) inasmuch as it is an end, i.e., that for the sake of which something is done; for example, health is the cause of walking. For if we are asked why someone took a walk, we answer, "in order to be healthy"; and in saying this we think we have given the cause. And whatever occurs on the way to the end under the motion of something else is also a cause. For example, reducing, purging, drugs and instruments are causes of health; for all of these exist for the sake of the end, although they differ from each other inasmuch as some are instruments and others are processes."

— Metaphysics 1013a, trans. John P. Rowan, Chicago, 1961

- Refers to the material cause, what a being's matter consists of (if applicable).
- Refers to the formal cause, what a being's essence is.
- Refers to the efficient cause, what brings about the beginning of, or change to, a being.
- Refers to the final cause, what a being's purpose is.

Unlike many ancient Greeks, who thought that an infinite regress of causality is possible, Aquinas argues that an infinite chain never accomplishes its objective and is thus impossible. Hence, a first cause is necessary for the existence of anything to be possible. Further, the First Cause must continuously be in action (similar to how there must always be a first chain in a chain link), otherwise the series collapses:

The Philosopher says (Metaph. ii, 2) that "to suppose a thing to be indefinite is to deny that it is good." But the good is that which has the nature of an end. Therefore it is contrary to the nature of an end to proceed indefinitely. Therefore it is necessary to fix one last end.

— Summa, II-I, Q.1, art.4.

Thus, both Aristotle and Aquinas conclude that there must be an uncaused Primary Mover, because an infinite regress is impossible.

However, the First Cause does not necessarily have to be temporally the first. Thus, the question of whether or not the universe can be imagined as eternal was fiercely debated in the Middle Ages. The University of Paris's condemnation of 1270 denounced the belief that the world is eternal. Aquinas' intellectual rival, Bonaventure, held that the temporality of the universe is demonstrable by reason. Aquinas' position was that the temporality of the world is an article of faith, and not demonstrable by reason; one could reasonably conclude either that the universe is temporal or that it is eternal.

Goodness

As per the Nicomachean Ethics of Aristotle, Aquinas defines "the good" as what all things strive for. E.g., a cutting knife is said to be good if it is effective at its function, cutting. As all things have a function/final cause, all real things are good. Consequently, evil is nothing but privatio boni, or "lack of good", as Augustine of Hippo defined it.

Dionysius says (Div. Nom. iv), 'Evil is neither a being nor a good.' I answer that, one opposite is known through the other, as darkness is known through light. Hence also what evil is must be known from the nature of good. Now, we have said above that good is everything appetible; and thus, since every nature desires its own being and its own perfection, it must be said also that the being and the perfection of any nature is good. Hence it cannot be that evil signifies being, or any form or nature. Therefore it must be that by the name of evil is signified the absence of good. And this is what is meant by saying that 'evil is neither a being nor a good.' For since being, as such, is good, the absence of one implies the absence of the other.

—Summa, I, Q.48, art.1.

Commentating on the aforementioned, Aquinas says that "there is no problem from the fact that some men desire evil. For they desire evil only under the aspect of good, that is, insofar as they think it good. Hence their intention primarily aims at the good and only incidentally touches on the evil."

As God is the ultimate end of all things, God is by essence goodness itself. Furthermore, since love is "to wish the good of another," true love in Thomism is to lead another to God. Hence why John the Evangelist says, "Whoever is without love does not know God, for God is love."

5.7 EXISTENCE OF GOD

Thomas Aquinas holds that the existence of God can be demonstrated by reason, a view that is taught by the Catholic Church. The quinque viae (Latin: five ways) found in the Summa Theologica are five possible ways of demonstrating the existence of God, which today are categorized as:

1. Argumentum ex motu, or the argument of the unmoved mover;
2. Argumentum ex ratione causae efficientis, or the argument of the first cause;
3. Argumentum ex contingentia, or the argument from contingency;
4. Argumentum ex gradu, or the argument from degree; and
5. Argumentum ex fine, or the teleological argument.

Despite this, Aquinas also thought that sacred mysteries such as the Trinity could only be obtained through revelation; though these truths cannot contradict reason:

The existence of God and other like truths about God, which can be known by natural reason, are not articles of faith, but are preambles to the articles; for faith presupposes natural knowledge, even as grace presupposes nature, and perfection supposes something that can be perfected. Nevertheless, there is nothing to prevent a man, who cannot grasp a proof, accepting, as a matter of faith, something which in itself is capable of being scientifically known and demonstrated.

—Summa, I, Q.2, art.2.

Aquinas responds to the problem of evil by saying that God allows evil to exist that good may come of it, but does not personally cause evil Himself.

5.7.1 View of God

Aquinas articulated and defended, both as a philosopher and a theologian, the orthodox Christian view of God. God is the sole being whose existence is the same as His essence: "what subsists in God is His existence." Consequently, God cannot be a body, He cannot have any accidents, and He must be simple. Further, He is goodness itself, perfect, infinite, omnipotent, omniscient, happiness itself, knowledge itself, love itself, omnipresent, immutable, and eternal. Summing up these properties, Aquinas offers the term actus purus.

Aquinas held that not only does God have knowledge of everything, but that God has "the most perfect knowledge," and that it is also true to say that God "is" His understanding.

Aquinas also understands God as the transcendent cause of the universe, the "first Cause of all things, exceeding all things caused by Him," the source of all creaturely being and the cause of every other cause. Consequently, God's causality is not like the causality of any other causes, because He is the transcendent source of all being, causing and sustaining every other existing thing at every instant. Consequently, God's causality is never in competition with the causality of creatures; rather, God even causes some things through the causality of creatures.

Aquinas was an advocate of the "analogical way", which says that because God is infinite, people can only speak of God by analogy, for some of the aspects of the divine nature are hidden (Deus absconditus) and others revealed (Deus revelatus) to finite human minds. Thomist philosophy holds that we can know about God through his creation (general revelation), but only in an analogous manner. For instance, we can speak of God's goodness only by understanding that goodness as applied to humans is similar to, but not identical with, the goodness of God. Further, he argues that sacred scripture employs figurative language: "Now it is natural to man to attain to intellectual truths through sensible objects, because all our knowledge originates from sense. Hence in Holy Writ, spiritual truths are fittingly taught under the likeness of material things."

In order to demonstrate God's creative power, Aquinas says: "If a being participates, to a certain degree, in an 'accident,' this accidental property must have been communicated to it by a cause which possesses it essentially. Thus iron becomes incandescent by the action of fire. Now, God is His own power which subsists by itself. The being which subsists by itself is necessarily one."

6

CARTESIANISM

6.1 INTRODUCTION

René Descartes worked out his metaphysics at a time of rapid advance in human understanding of the physical world. He adopted from Galileo the view that physical things are not what they are commonly taken to be on the strength of sense experience—namely, possessors of "secondary" properties such as colour, smell, and texture—but rather are objects characterized only by the "primary" qualities of shape, size, mass, and mobility. To understand why a constituent of the physical world behaves as it does, what should be asked is: Where is it? How large is it? In what direction is it moving, and at what speed? Once these questions are answered, its further properties will become intelligible. Descartes held further that all change and movement in the physical world is to be explained in purely mechanical terms. God was needed to give initial impetus to the physical system as a whole, but, once it had got going, it proceeded of its own accord. To pretend, as the Aristotelians had, to discern purposes in nature was to make the impious claim to insight into God's mind. Descartes applied this theory to the movements of animals as much as to those of inanimate bodies; he thought of both as mere automatons, pushed and pulled about by forces over which they had no control.

Although Descartes thus acquiesced in—indeed, emphasized—the mechanistic tendencies of contemporary science, he was far from being a materialist. In his Meditationes de Prima Philosophia, he argued that there was a total and absolute distinction between mental and material substance. The defining characteristic of matter was to occupy space, while the defining characteristic of mind was to be conscious or, in a broad sense of the term, to think. Material substance was, so to speak, all one, although packets of it were more or less persistent. Mental substance existed in the form of individual minds, with God as the supreme example. The mental and material orders were each complete in themselves, under God; it was this assumption that made it appropriate for Descartes to use the technical term substance in this context.

The main difficulty faced by Descartes's mind-body dualism was that of bringing together the two orders of being once they were separated. Some later Cartesians inferred from the fact of this separation that there can be no interaction between mind and body: all causality is immanent, within one order or the other, and any appearance of mind affecting body or of body affecting mind must be explained as the result of a special intervention by God, who, on the occasion of changes in one substance, brings it about that there are corresponding changes in the other. Descartes himself, however, had no sympathy with this view, which was called occasionalism.

6.2 KANTIANISM

It is worth mentioning here another move in the same area that many have found instructive. Immanuel Kant, who was in some respects both a latter-day Cartesian and a latter-day Platonist, argued that human activities could be looked at from two points of view. From the theoretical standpoint, they were simply a set of happenings, brought about by antecedent events in precisely the same way as occurrences in the natural world. From the standpoint of the agent, however, they must be conceived as the product of rational decision, as acts for which the agent could be held responsible. The moment people begin to act, they transfer themselves in thought from the phenomenal world of science to an intelligible world of pure spirit; they necessarily act as if they were not determined by natural forces. The transference, however, was a transference in thought only, and because of this the problem of the unity of the universe was dissolved. There was no contradiction in people thinking of themselves both as subjects for science and as free originators of action. Contradiction would appear only if they were present in both respects in an identical capacity. But appeal to the doctrine of the two standpoints was thought by Kant to rule this out.

It is only with some hesitation that one can speak of Kant as having put forward a metaphysics. He was in general highly suspicious of claims to metaphysical knowledge, and a principal aim of his philosophy was to expose the confusions into which professing metaphysicians had fallen. Nevertheless, it is clear that Kant had metaphysical convictions, for all his denial of the possibility of metaphysical knowledge; he was committed to the view that humans can conceive a nonnatural as well as a natural order and must necessarily take the former to be real when they act.

6.2.1 Idealism

Descartes and Kant were both adherents of metaphysical dualism, though they worked out their dualisms in interestingly different ways. Many thinkers, however, find dualism unsatisfactory in itself; they look for a single principle to encompass whatever exists. There are two broad steps that are open to those who confront a dualism of mind and matter (a mind-body dualism) and find it unsatisfactory: they can either try to show that matter is in some sense reducible to mind or, conversely, seek to reduce mind to matter. The first is the solution of idealism, the second that of some versions of materialism.

There are various forms of idealism. In one version, this philosophy maintains that there literally is no such thing as matter; what the common person takes to be material things are, upon closer consideration, nothing but experiences in minds. Nothing exists but minds and their contents; an independently existing material world is strictly no more than an illusion. This was the view taken by George Berkeley. In the more sophisticated idealism of G.W.F. Hegel, however, it is not maintained that mind alone exists; material things are, in one way, taken to be as real as minds. The thesis advanced is rather that the universe must be seen as penetrated by mind—indeed, as constituted by it. "Spirit," to use Hegel's own word, is the fundamental reality, and everything that exists must accordingly be understood by reference to it, either as being directly explicable in spiritual terms or as prefiguring or pointing forward to spirit.

Whatever the merits of this thesis, it is clear that it differs radically from that maintained by Berkeley. Idealism as Berkeley espoused it relies largely on arguments drawn from epistemology, though formally its conclusions are ontological, because they take the form of assertions or denials of existence. Hegel, however, had little or nothing to say about epistemology and was not even concerned to put forward an ontology. What he wanted to urge was a doctrine of first principles, a thesis about the terms in which to understand the world. The Hegelian "reduction" of matter to mind was thus reduction in a somewhat attenuated sense.

6.3 MATERIALISM

One of the first questions to engage the attention of the ancient Greek philosophers was "What is the world made of?". Of the many answers they proposed, one became increasingly influential and is widely accepted today: matter.

The view that the world is entirely material, consisting entirely of matter, is called materialism. It is one of three monistic theories of the nature of the world, the other two being idealism and neutral monism. (According to the latter view, the world consists entirely of some fundamental stuff or substrate that is neither mental nor material; what are called the mental and the material are in some sense aspects or manifestations of this fundamental "neutral" reality.)

It is often convenient to characterize materialism negatively, in terms of the metaphysical theories to which it is opposed. Such theories include not only the other two monistic theories just mentioned but also any dualistic theory according to which the world is partly—but only partly—material. One example of such a dualistic view is supernaturalism, which holds that there are, in addition to material things, certain immaterial things—conscious, purposeful, rational beings—whose existence is independent of and prior to the existence of matter and whose natures are in some sense superior to the natures of merely material things. Some supernaturalists believe that human beings are wholly material things and, therefore, not immaterial beings.

Although such philosophers are for this reason sometimes called "materialists," they are not materialists in the strict sense of the term. A better-known example is, of course, mind-body dualism, which was dominant in the philosophy of mind from the time of Descartes to about the early 20th century.

Materialism rejects the two main kinds of mind-body dualism: substance dualism, or psychophysical dualism—the view held by Descartes—and property dualism. According to the latter, thinkers are material things but nevertheless have properties—such as "being in pain" or "believing that Tokyo is the capital of Japan"—that are not material or physical. These properties are said to be not material or physical in the sense that their content does not consist of a specification of interactions among the material or physical parts of the things to which the properties belong. Physiological and neurological properties—for example, "has among its constituent cells neurons that send one another signals propagated by charge-carrying ions," a property of the brain—do consist of such specifications. Materialism denies both that there are immaterial substances and that there are properties other than material or physical properties.

Materialistic philosophies of the mental, of thought and sensation, are themselves of two kinds. According to reductive materialism also called the identity theory, thought and sensation are real but involve nothing that is in any sense immaterial. It may be, for example, that a given person's pain sensation or thought that it is about to rain is identical with—is literally one and the same thing as—a certain complex interaction among some of the atoms that compose that person's brain. According to eliminative materialism, there are, in reality, no thoughts or sensations at all. Eliminative materialism holds that the phenomena that have been described and accounted for by reference to thought and sensation should be described and accounted for entirely in terms of behaviour and its physiological causes.

It is also customary to distinguish two kinds of reductive materialism: token materialism and type materialism. Consider, for example, the phenomenon of pain. Each particular episode of pain is said to be a token of the type "pain." Token materialists hold that each particular episode of pain is identical to some particular physical event. Type materialists hold that pain—the general phenomenon—is identical to some general physical phenomenon, such as the firing of C fibres in an organism's brain. Token materialists who reject type materialism point out that it seems possible for there to be two organisms that experience pain but have entirely different physiological structures.

In the second half of the 20th century, many analytic philosophers came to regard the concept of matter as having been rendered problematic by the advent of quantum mechanics in modern physics. One would suppose, for example, that if a "material" thing has parts, all its parts must be material things. And it would seem that every material thing must occupy a definite region of space. But a chair is a material thing if anything is, and, according to physics, every chair has electrons as parts. Yet, according to quantum mechanics, an unobserved electron cannot be said to be located in any definite region of space.)Such philosophers, accordingly, came to prefer the term physicalism or naturalism to materialism, though the latter term continued to be generally accepted.

6.3.1 Naturalism

There are two varieties of naturalism: ontological and epistemological. Ontological naturalism is the thesis that everything that exists is a part or aspect of the natural world. More or less equivalent formulations of the thesis would be: "Everything that exists is a part or aspect of the physical world," "...of nature," "...of the cosmos," "...of the physical world," and "...of the physical universe." The word physical is derived from the Greek physis, usually translated as "nature"; natural and nature come from its Latin counterpart, natura.)

Ontological naturalism is thus much the same thesis as materialism or physicalism. There is, however, a rhetorical difference between the terms materialism and physicalism on the one hand and the term naturalism on the other. Materialism and physicalism are used in opposition to the three terms supernaturalism, idealism, and dualism, and naturalism is used in opposition to supernaturalism and nonnaturalism. Nonnaturalism is simply the denial of naturalism: it is the thesis that there are parts or aspects of reality that cannot be described as parts or aspects of nature or the physical world, but these "nonnatural" parts or aspects of reality are not necessarily supernatural, not necessarily above or causally antecedent to nature. G.E. Moore's ethical intuitionism and property dualism in the philosophy of mind are examples of nonnaturalistic theories that are not supernaturalistic. The rhetorical difference between materialism/physicalism and naturalism can be compared to the difference between theism and monotheism: the latter two are names for the same doctrine, but theism is used in opposition to atheism and monotheism in opposition to polytheism.

Epistemological naturalism takes no stand on whether supernatural or nonnatural things exist but insists that if such entities do exist, it is impossible to have any knowledge of them, that belief in them is unwarranted, and that speculation about them is pointless.

There is a strong tendency among naturalists of both varieties to explain the concept of "nature" by reference to the natural or physical sciences: nature is a collective name for those things that are investigated, revealed, or postulated by the natural sciences. Understood in those terms, ontological naturalism is the thesis that the things whose existence is asserted or postulated by the sciences constitute the whole of reality. And epistemological naturalism is the thesis that one should believe in the existence of only those things whose existence is asserted or postulated by the sciences. It is, moreover, generally held by naturalists that everything asserted to exist by any science must in some sense be composed of or reducible to the fundamental entities of physics.

Naturalists recognize that physics postulates different fundamental entities at different times. Naturalists are, however, convinced that, however much the properties ascribed to the fundamental entities of physics may vary over time, those entities will always have the following feature: their properties and mutual relations will be quantifiable and will in no sense be mental or teleological. In other words, such entities will have only properties that are like charge and mass one particle may have twice the charge or half the mass of another and stand to one another only in relations that are like spatial distance the distance between two particles must be some multiple of the distance between two other particles. Ontological naturalism, then, may be viewed as the doctrine that everything is ultimately composed of things all of whose properties are quantifiable and none of whose properties is in any way mental or teleological. The doctrine also maintains that interactions among such things are entirely determined by one all-encompassing set of physical laws that refer only to quantifiable properties and relations. If anything has either mental or teleological properties, it will be a large, contingent structure such as a living organism that is ultimately composed of things that lack such properties, and all of the properties of that structure will be determined by the properties of its ultimate parts and the causal relations that hold among them.

The foregoing statement of ontological naturalism marks its fundamental difference from supernaturalism, according to which mentality and teleology are fundamental features of reality. According to supernaturalism, there are conscious, purposeful beings God or gods or spirits of some sort whose mental and teleological properties are not consequences of their having nonconscious, nonpurposeful parts that existed before they did and are currently arranged in a way that generates thought and purpose.

Epistemological naturalism, finally, maintains that if there are entities that are not ultimately composed of nonconscious, nonpurposeful parts, one can have no reason to believe in their existence, and speculation about them is pointless.

6.4 CONTEMPORARY METAPHYSICS

6.4.1 Analytic Metaphysics in the 20th Century

The 20th century was not kind to metaphysics. The middle years of the century witnessed the ascendance of philosophical movements that dismissed all metaphysical questions as meaningless or confused. The tradition that has come to be known as "analytic philosophy" was dominated by those anti-metaphysical forces for roughly 30 years, beginning in the mid-1930s, when the English philosopher Sir A.J. Ayer (1910–89) brought the logical positivism of the Vienna Circle to Britain. During the same period, opponents of analytic philosophy in Anglophone philosophy departments came to be called "Continental philosophers" because of their greater interest in philosophical developments in continental Europe. Their tradition proved no more hospitable to metaphysics, however. Although Continental philosophers are heterogeneous, representing radically different doctrines and ways of doing philosophy, upon one thing they have tended to agree: Kant's criticisms of "dogmatic" metaphysics—which attempted to arrive, through reason alone, at truths about objects beyond any human experience—are unanswerable. Later in the 20th century, however, metaphysics made a spectacular recovery within the analytic tradition.

Use of the term philosophy of analysis can be traced back to the first decade of the century, when the English philosophers Bertrand Russell (1872–1970) and G.E. Moore (1873–1958) began to criticize the idealism typified by the work of the English philosopher F.H. Bradley (1846–1924)—a blow from which idealism never recovered. The battle with the idealists included a disagreement over the feasibility of the "analysis of facts." According to Bradley, any attempt to break down a fact into its components is doomed to failure; facts are indivisible, unanalyzable wholes. Russell and other opponents of idealism, such as the American "new realists" Walter T. Marvin (1872–1944) and Ralph Barton Perry (1876–1957), maintained that the components of facts can be identified by analysis. Even though the original fact would admittedly not appear on a mere list of its components, that need not imply that the listed entities are not, after all, components. By mid-century, when philosophers talked of analysis, they generally meant conceptual or linguistic analysis. But, as used by the original analytic philosophers, analysis was the name of a nonlinguistic activity: it was the prying apart (in thought) of the very contents of the world—a delicate metaphysical procedure.

Russell—with his collaborator, the English philosopher Alfred North Whitehead (1861–1947)—had built upon the advances in mathematical logic made by the German mathematician and philosopher Gottlob Frege (1848–1925), and respect for logical rigour has proved to be one of the few constants in the analytic tradition. As the century progressed, the label analytic philosophy was retained but applied to increasingly various methods and doctrines, many of which had little in common with those advocated by Russell and Moore. Perhaps the most one can say about what unites philosophers in the analytic tradition is that they would almost invariably regard Russell and Moore as the heroes in the battle with the British idealists.

During the first half of the 20th century, the primary aim of analytic metaphysics was to articulate and defend realism. The idealists had held that anything of which the mind could be aware must itself be mental or mind-dependent in some respect. The initial realist reaction was extreme, as Moore and some of the new realists denied the mind-dependence of virtually everything of which the mind could be aware, a position that was difficult to maintain. Moore, Russell, and many other early 20th-century realists took perception to involve awareness of "sense data." Whenever it appears to someone that there is something round or red, there is, they thought, a round or red sense datum of which that person is aware. That position raised difficult questions about the relationship between sense-data and the physical objects they somehow disclose to perceivers, and it called into question the more extreme forms of realism. For example, although no real (physical) bloody dagger appears to the victim of hallucination, there is an apparent dagger. Therefore, on the sense-datum theory, there must be a red triangular sense-datum that the victim is experiencing. Do not such "wild" sense-data, at least, belong to the realm of the mind-dependent? What of illusions, as when a straight stick in water appears bent, or white paper under red light appears red? If the sense-data are bent and red but no physical objects in the neighbourhood are bent or red, then such sense-data, at least, would seem to be mental entities.

The braver realists took wild sense-data to be as external to the subject as more mundane sense-data. Some American new realists explained illusion by a metaphysical tour-de-force: they built physical objects out of all the appearances that such objects could possibly present, construing perception as the selection of some of those appearances, not as the occasion of their generation. In illusion, sense-data are selected that really do belong to an object, and their illusoriness consists only in the fact that they tend to mislead the subject about the object's other, unexperienced appearances.

American philosophers who called themselves "critical realists" held a range of alternative views. Durant Drake (1878–1933) and Charles Augustus Strong (1862–1940) took veridical perception to involve a relation of awareness holding between a perceiver and a property that is exemplified by some physical object. In illusion and hallucination, they said, the relation of awareness holds between a person and the very same property, even though the property is not exemplified by anything. The critical realism of American philosopher Arthur O. Lovejoy (1873–1962) was more straightforwardly dualistic: although sense-data are mental entities, the world that is known by means of them is not.

Russell defended a dualism of this kind in The Problems of Philosophy (1911), but he eventually came to an "under-the-hat" theory: sense-data are physical processes in the brain. Russell drew the conclusion that one sees only those things that happen inside one's own head. Roy Wood Sellars (1880–1973), a Canadian-born critical realist, agreed that sense-data are brain processes but denied that they are seen. Sense-data, for him, were brain processes by means of which other things are seen.

Sellars's view fits neatly with the adverbial theory of sensation espoused by the American philosopher C.J. Ducasse (1881–1969), according to which the experience of a blue patch of colour, whether veridical or illusory, is not properly taken to be a relation between a subject and something blue. To dance a waltz, for example, is just to dance "waltzily," not to stand in the dancing relation to some further thing distinct from the dance. Similarly, to have a blue sense-datum is just to sense "bluely," or to undergo a sensory experience of a distinctive type. It is not a matter of being related to a peculiar kind of entity that can itself be said to be blue. The American philosophers Roderick Chisholm (1916–99) and Wilfrid Sellars (1912–89), the son of Roy Wood Sellars, were also proponents of adverbialism during the second half of the century.

The attempt to develop a metaphysics of sensory experience arguably reached its most sophisticated form in Perception (1932), by the English philosopher H.H. Price

(1899–1985). By then, however, the anti-metaphysical waves were about to break upon the entire generation of philosophers of which Price was a member, and their work was subsequently forgotten. By the time metaphysics recovered some of its lustre in the 1960s, only Russell and Moore were still famous figures—Russell being lauded chiefly for his theory of definite descriptions and his contributions to logic and Moore being rehabilitated as an "ordinary language" philosopher avant le mot (i.e., before that term had come into common usage). However, in the last decade of the century, the Australian philosopher David Chalmers (born 1966) popularized a kind of "property dualism" about qualia—the "what-it-is-like" aspect of experience, which the American philosopher Thomas Nagel (born 1937) had put back on the agenda for philosophers of mind in a famous article, "What Is It Like to Be a Bat?" (1974). Property dualism—the view that mental properties are nonphysical properties and, hence, that the physical properties of an organism do not determine its mental properties—demanded answers to all questions about the status of sense-data. As a result, the metaphysical programs of the early realists and critical realists no longer seemed as quaint as they did mid-century.

Russell and Moore had understood that their views were closer to the realisms of the Austrian philosophers Franz Brentano (1838–1917) and Alexius Meinong (1853–1920) than to the idealisms that were popular in Britain and North America. But Meinong's unbridled realism about all objects of thought—including nonexistent "golden mountains" and even impossible objects, such as the "round square"—was too much for Russell's "robust sense of reality," which he insisted was necessary "in the analysis of propositions about unicorns, golden mountains, round squares, and other such pseudo-objects". Russell's theory of definite descriptions had shown that empty singular terms— e.g., the present king of France, the golden mountain—could be used in a way that did not commit the speaker to a shadowy kind of nonexistent entity referred to by those terms. To say that there is no such thing as the golden mountain is not to attribute nonexistence to an entity that must, then, have some lower grade of being; it is rather to say that there is nothing which is uniquely a mountain made of gold. Quine's criterion of ontological commitment (a scientific theory, as expressed in a formal language such as predicate logic, is committed to all and only those entities that must exist if the theory is to be true), did much to restore the fortunes of metaphysical theorizing in the 1950s. The criterion was a natural extension of Russell's idea that, to find out what a theory really commits one to, singular terms should be removed in favour of quantifier phrases, such as some, all, and none.

Quine's method seemed to provide a principled way to determine what kinds of thing philosophers must accept as real, given the theories that they hold. Quine himself served as a positive test case for the method's efficacy. Empiricist inclinations had drawn him to a strict nominalism that would regard even the existence of sets (or classes) as excessively "Platonistic." Finding that the theories of mathematical physics implied the existence of mathematical objects—objects such as functions of real and complex variables—Quine found himself a reluctant convert to Platonism. He opted, however, for the most austere form of Platonism possible, one that recognized no nonindividuals other than sets.

Quine cared primarily about the ontological commitments of the physical sciences. What is known, he thought, is restricted to the domains of those sciences and mathematics; all the rest is mere opinion, and there is no point in attempting to achieve clarity about the ontological commitments of "theories" that amount to little more than imprecise guesses. Naturally enough, less skeptical metaphysicians adopted the Quinean method, attempting to make their ontological commitments clearer by proposing paraphrases of the contents of much broader bodies of belief than Quine would take seriously. Soon the metaphysical case files were reopened on the existence of universals, propositions, events, holes, merely possible objects, and many other kinds of entities—not just Quine's sets.

Quine, like Russell before him, was skeptical about the part of metaphysics concerned with modal notions: necessity, possibility, and the essential properties of things. Modality was, however, the next traditional metaphysical topic to witness a dramatic revival. Early in the century, the American philosopher C.I. Lewis (1883–1964) had developed several systems of modal logic, but it was hard to see any principled way to choose between them when looking for the logical rules governing the most general kinds of necessity and possibility. The American philosopher Ruth Barcan Marcus (1921–2012) and the German-born American logician Rudolf Carnap (1891–1970) independently developed versions of quantified modal logic in 1946. Unlike C.I. Lewis's propositional modal logics, their systems allowed for the formal analogues of sentences like Necessarily, all cyclists have two legs and All cyclists necessarily have two legs. Although Carnap's system treated the two as equivalent, it is natural to take the latter, but not the former, as meaning that every person who happens to ride a bicycle is essentially two-legged.

Quine was suspicious of the notion of essence required to make such a distinction, but his arguments against reading necessarily, in this context, as meaning "essentially" were refuted by the American mathematician Raymond Smullyan (1919–2017) in 1948. Finally, in the 1960s and '70s necessity, possibility, and essences became central to metaphysics in a way that they had not been since the days of the great rationalist philosophers Benedict de Spinoza (1632–77) and G.W. Leibniz (1646–1716).

The American philosopher Saul Kripke (born 1940), while still a teenager, made important contributions to the formal side of the study of modality. In his semantics for modal logic, Carnap had made use of "state descriptions," which were intended to play the role of Leibniz's possible worlds. Kripke's semantics for modal logic took the notion of a possible world for granted and showed how to interpret the differences between C.I. Lewis's modal logics in a systematic way. The result was a much clearer understanding of what was at stake in the choice of principles governing necessity and possibility. In the 1970s Kripke argued convincingly that, when scientists discover that a natural kind of substance (e.g., water) has a certain underlying nature (e.g., consisting of H_2O molecules), what they have discovered is an essential property of the kind in question. There are, then, necessary truths best investigated using methods that are a posteriori—i.e., derived from or based on experience. He also boldly ascribed essential properties to individuals (e.g., people necessarily have the parents they actually have) and paid no heed to the anti-essentialist prejudices of previous generations of philosophers.

By the 1980s nearly all the traditional problems of metaphysics were once again the subject of intense debate within analytic circles. Ontological and cosmological arguments for the existence of God were defended by prominent philosophers, as was the compatibility of determinism and free will. Almost every traditional metaphysical school—including forms of idealism, Thomism, and rationalism—had representatives operating squarely within the analytic movement. A list of the most prominent analytic philosophers of the second half of the 20th century would include a high proportion who would become renowned, in large part, for their contributions to metaphysics: e.g., from the United States, Willard Van Orman Quine, Saul Kripke, Nelson Goodman (1906–98), Wilfrid Sellars, Roderick Chisholm, Donald Davidson (1917–2003), Hilary Putnam (1926–2016), Alvin Plantinga (born 1932), Sydney Shoemaker (born 1931), David Lewis (1941–2001), and Peter van Inwagen (born 1942); from Australia, J.J.C. Smart (1920–2012) and David Armstrong (1926–2014); and, from the United Kingdom, Sir Peter Strawson (1919–2006) and Derek Parfit (1942–2017). Thus, by the turn of the millennium, metaphysics in the analytic tradition had largely regained the prominence and prestige it had lost during the middle years of the 20th century.

6.4.2 Continental Metaphysics in the 20th Century

In the 20th century, metaphysics was faced with an especially difficult challenge: how to restore and perpetuate the mission of what Aristotle had called "first philosophy" following the advent of Darwinism and positivism, which had presented plausible scientific alternatives to traditional religious and metaphysical accounts of human origins, human nature, and human society. In the wake of those developments and for much of the 20th century, the very use of the term metaphysics could suggest an anachronistic viewpoint. As the German philosopher Theodor Adorno (1903–69) observed in his lectures on metaphysics, delivered in 1965 and posthumously published as Metaphysik: Begriff und Probleme, "Today metaphysics is used in almost the entire non-German-speaking world as a term of abuse, a synonym for idle speculation, mere nonsense and heaven knows what other intellectual vices." That situation illustrated one of the central dilemmas of 20th-century Continental philosophy: it wished to distance itself from inherited metaphysical paradigms, which seemed epistemologically naive, yet, in doing so, it could not entirely renounce them, for fear of ceding too much ground to naturalism and scientism and thereby surrendering its raison d'être.

The foremost representatives of this dilemma may be found among proponents of phenomenology (a method of philosophizing aimed at the direct investigation and description of the phenomena of conscious experience), particularly Edmund Husserl (1859–1938), its founder, and Martin Heidegger (1889–1976). If one understands metaphysics as the search for irrefutable, a priori first principles on the basis of which Being in its totality, can be explained, phenomenology is, strictly speaking, anti-metaphysical. By the same token, since its inception in the work of Husserl, phenomenology has strived to reconceptualize traditional metaphysical questions about the fundamental nature of reality in a non-metaphysical way and on the basis of rigorously obtained evidence. For Husserl, that basis was constructed through the doctrine of intentionality: the idea that, because consciousness is always "consciousness of" something, the radical epistemological separation between subject and object, so characteristic of the philosophies of René Descartes (1596–1650) and Immanuel Kant (1724–1804), is fundamentally mistaken. The assumed separation presupposes as primary a cognitive problem that has not been demonstrated to exist.

In his essay Philosophie als strenge Wissenschaft and other works, Husserl was at pains to show that transcendental phenomenology can reestablish the traditional goals of first philosophy on an immanent, nonspeculative basis. Husserl thereby suggested that, through the phenomenological approach, one could arrive at truths about the nature of Being or of experience that are universal, necessary, and certain—a goal that sounded extremely similar to the objective of traditional metaphysics. Yet, with phenomenology one no longer begins with principles that are a priori, such as the forms of Plato, the unmoved mover of Aristotle (384–322 BCE), or the absolute Spirit (Geist) of G.W.F. Hegel (1770–1831). Instead, one arrives at ontological truth via the phenomenological-intentional experience of phenomena, or "things themselves"—hence Husserl's deceptively simple slogan or watchword, "to the things" . Thus, for Husserl and his followers, the metaphysical quest for truth must be reconceived as an immanent this-worldly investigation. The object of Husserl's theory of Wesensschau or the intuition of essences, as set forth in Ideen zu einer reinen Phänomenologie und phänomenologischen Philosophie (1913; Ideas: General Introduction to Pure Phenomenology) and other works, is to indicate the invariant structures of human experience: what it means to perceive an object, what it means to formulate a judgment, what it means to explore the nature of temporality, and what it means to probe the depths of human consciousness.

Heidegger, Husserl's onetime assistant at the University of Freiburg, similarly took the phenomenological method as his point of departure, though the results he obtained were markedly different from those of Husserl. Whereas the abstract nature of intentional consciousness, as Husserl conceived it, bears marked affinities to Kant's transcendental standpoint (i.e., the structure of possible experience), in Sein und Zeit Heidegger recast intentional consciousness as Dasein ("being-there") and insisted above all on its embodied nature. Thus, Dasein bears few resemblances to the narrowly cognitive, transcendental ego of Cartesianist epistemology. Instead, it is characterized by the priority of what Heidegger called "Being-in-the-world," aspects of which are moods, anxiety, idle talk, the "they" , "falling," "Being-with-others," and so forth. Such "existentials", or fundamental structures, of Heidegger's existential ontology are the experiential prisms through which Dasein interacts with the world.

By the same token, in Being and Time Heidegger also insisted that the analytic (analysis) of Dasein is a necessary prelude or propaedeutic to exploration of the Seinsfrage, or the question of the meaning of Being. It was in that spirit that Heidegger frequently acknowledged the seminal value for his thought of Aristotle's celebrated dictum, "Being is said in many ways"—i.e., there are many senses of the word being. Hence, from the outset of his philosophical career—as far back as his dissertation on Duns Scotus (c. 1266–1308)—Heidegger, under Aristotle's influence, was keenly interested in questions about the nature of Being, a fact which suggests that he regarded metaphysics more positively than Husserl had.

Indeed, throughout Heidegger's early writings—roughly, until the lectures on Friedrich Nietzsche (1844–1900) in 1936–40—the idea of metaphysics played an essentially positive role in his philosophy. For Heidegger during that period, to engage in metaphysics meant to address the Seinsfrage in phenomenological terms, or on the basis of his own preferred approach, "fundamental ontology." The history of philosophy has known numerous inquiries into the nature of individual beings. However, what has been glossed over or covered up is what is most fundamental or primordial: the nature of Being itself. For Heidegger, human beings' metaphysical relationship to Being structures their entire existence. If they get the answer to this question wrong, they will fall short in every other respect as well.

A brief survey of some of the key works of Heidegger's early philosophy reveals metaphysics's centrality to the project of fundamental ontology. Crucial in that respect was his 1929 Freiburg inaugural lecture, appropriately titled "What Is Metaphysics?" In retrospect, the address appears as an important way station en route to the so-called Kehre ("Turn") in Heidegger's thought: the transition from a Dasein-centred existential paradigm to a perspective that emphasized the standpoint of Being itself. The "forgetting" of the Seinsfrage, insisted Heidegger, makes the human world ontologically poorer. During that phase, Heidegger viewed metaphysics, when properly reconceived, as an important ally in the reformulation of the Seinsfrage from the standpoint of Being.

The leitmotif of the Freiburg inaugural lecture was das Nichts, or "the Nothing." According to Heidegger, the main shortcoming of metaphysics throughout its history is that it has been preoccupied with individual beings. It has therefore neglected, to its own detriment, that status of non-Being, or nothing. "How is it with the Nothing?" Heidegger inquires with false naiveté, thereby seeking to certify it as a legitimate topic of metaphysical inquiry. We intuit the Nothing in the mood of fundamental "anxiety", which calls into question not this or that particular aspect of existence but existence as a whole. "Anxiety robs us of speech," observed Heidegger. "Because beings as a whole slip away, so that just the Nothing crowds round, in the face of anxiety all utterance of the 'is' falls silent."

In a celebrated formulation, Heidegger declared that das Nichts nichtet: "the Nothing nothings." Traditional ontology has focused on the nature of Being. But it has woefully neglected non-Being, or Nothing, which plays a crucial role in what Heidegger referred to as the "concealment of Being". Heidegger claims that every unveiling of Being is simultaneously an act of concealment. The dialectical interplay of concealment and unconcealment defines the way in which Being manifests itself or comes to presence. In a famous discussion of What Is Metaphysics?, "Überwindung der Metaphysik durch die logische Analyse der Sprache", the logical-positivist philosopher Rudolf Carnap seized on Heidegger's claim that "the Nothing nothings" as proof that fundamental ontology, as a variant of metaphysics, is meaningless, or devoid of sense.

As further evidence of Heidegger's sympathy for metaphysics, one might invoke two complementary texts from the late 1920s and early '30s: Die Grundberiffe der Metaphysik, originally a series of lectures delivered in 1929–30, and Kant und das Problem der Metaphysik. And in 1935 Heidegger presented a lecture course, Einführung in die Metaphysik, in which he stressed the more primordial capacity for metaphysical questioning possessed by ancient Greek thought.

Heidegger's attitude toward metaphysics was always characterized by a distinct ambivalence. On the one hand, he felt that previous approaches to metaphysics were marred by an inferior and unproductive way of posing basic philosophical questions, such as Seinsfrage. It was in this vein that Heidegger, in Being and Time, recommended the "destruction" of all prior metaphysics. On the other hand, until the mid-1930s, he understood his own project as an effort to reestablish metaphysics on a sound philosophical basis, in keeping with the grandeur and sublimity of the Greek beginning.

As the 1930s wore on, however, Heidegger became extremely skeptical that there was anything in the metaphysical tradition worth salvaging. Consequently, after having been an advocate of metaphysics, he became one of its most thoroughgoing detractors. Increasingly, he viewed all of Western philosophy, beginning with Plato, as a falsification of the Seinsfrage as it had been originally formulated by pre-Socratic thinkers such as Heraclitus (c. 540–c. 480 BCE), Parmenides (born c. 515 BCE), and Anaximander (610–546 BCE). In Heidegger's view, Plato's doctrine of forms had already distorted the possibility of authentic philosophical questioning by stressing the primacy of "representation" and "ideas," as opposed to the concealment and unconcealment of Being itself. By the late 1930s, Heidegger's pessimism about metaphysics had become even more pronounced. In "Überwindung der Metaphysik", first published in 1954, for example, Heidegger equated the triumph of metaphysics with a condition of total nihilism. According to Heidegger, the metaphysical worldview that has predominated in the West has resulted in "the unconditional objectification of everything present." The "devastation of the earth" and the "collapse of the world," as he put it, follow from the fact that "metaphysical man" has achieved theoretical and practical supremacy.

One of the few significant contemporary Continental thinkers to repose metaphysical questions and themes unabashedly is the French philosopher Alain Badiou. He was also one of the few Continental philosophers after Heidegger to place the Seinsfrage at the centre of his philosophy. Like Aristotle in the Metaphysics, Badiou was especially interested in the apparent paradox that being, which is presumably One, presents itself in many ways: as necessary and as contingent, as fixed and as relational, as eternal and as transient. Accordingly, Badiou's central philosophical text, Being and Event (1988), opens with a meditation on the time-honoured ontological question concerning the relationship between the One and the Many, which is the theme of Plato's dialogue Parmenides. The Eleatic school, which included Parmenides and influenced Plato, vigorously insisted that Being is One . "If One is not," claimed Parmenides, "then nothing is." In contrast, Badiou, under the influence of post-structuralist thinkers such as Gilles Deleuze and Jacques Lacan, characterized Being as "infinite multiplicity," thereby siding firmly with the followers of Heraclitus, who believed that variation and change, rather than uniformity and stasis, were primary.

Badiou's other main theoretical influence was the set theory of the German mathematician Georg Cantor (1845–1918). As Badiou once declared: "Ontology is mathematics." In that respect too, Badiou's approach was reminiscent of Plato, who sought to legitimate philosophical truth by reference to the so-called "theory of numbers": the Pythagorean doctrine that unchanging mathematical axioms were the secret structural constants of Being. Badiou designated Cantor's set theory as the theoretical "event" of the 20th century. In Badiou's eyes, Cantor "laicized infinity" by demonstrating not only that sets are not comprehensive but also that they can never contain the totality of their members. Accordingly, the potential ways in which one might count the members of a set are infinite—hence Badiou's endorsement, following Cantor, of what Badiou called the "axiom of infinite multiplicity," which became the centrepiece of Badiou's ontology of pluralized indeterminacy. As with other post-structuralist approaches, Badiou's understanding of Being as "infinite multiplicity" had a practical aim: it sought to counteract metaphysical "closure" and render thought receptive to the eruption of the "Event"—an ontological breakthrough, or novum, which, according to Badiou, disrupts the limitations and constraints of Being qua "totality." Here Badiou's presumption is that to conceive of Being as pluralized indeterminacy as opposed to univocal oneness has an emancipatory effect. This is why Badiou claims that to conceptualize truth negatively as a "hole in knowledge" or as an "excess" vis-à-vis "exact designations" yields results that are ultimately positive. As Badiou argued in Manifeste pour la philosophie:

If truth makes a hole in knowledge, if there is hence no knowledge of truth, but only a production of truths, it is because mathematically thought in its being—thus as a pure multiplicity—a truth is generic, subtracted from exact designations, in excess with regard to what these designations afford to discern. The price to pay for this certainty is that the quantity of a multiple tolerates an indeterminacy, a kind of disjunctive rift that constitutes the whole of the real of being itself...

Thus, in Badiou's view, ontology can give rise only to "wandering excess," rather than to finality or completion.

6.5 CRITICISMS OF METAPHYSICS

6.5.1 The Epistemology of Metaphysics

When one considers the fact that metaphysicians—like all philosophers—have failed to reach a consensus on any important problem in their field, one is led naturally to the question of whether knowledge is possible in metaphysics. No answer to that question will satisfy all philosophers, but it is possible to provide a classification or taxonomy of possible answers to it.

All metaphysicians have affirmed metaphysical propositions of one sort or another and have, if only tacitly, claimed to be in possession of reasons for thinking those propositions to be true—reasons the possession of which would presumably be sufficient for knowledge. They have, however, only rarely explicitly addressed questions such as "If one knows a metaphysical proposition to be true, how does one know it to be true?" or "What is the source of metaphysical knowledge?" Examination of the writings of various metaphysicians, however, suggests various epistemologies of metaphysics, each of which is implicit in the metaphysician's approach to the questions and problems of metaphysics. Those epistemologies fall into the following four general categories.

Metaphysical propositions are known in the same way that mathematical propositions are known. That is, certain "basic truths of metaphysics" or "metaphysical axioms" are in some sense self-evidently true. And certain principles of reasoning are self-evidently valid. It is possible to use those principles of reasoning to deduce important and unobvious truths from the set of self-evident but indemonstrable basic truths. That conception of the basis of metaphysical knowledge was characteristic of the so-called Continental Rationalists—René Descartes (1596–1650), Benedict de Spinoza (1632–77), and Gottfried Wilhelm Leibniz (1646–1716). One way to summarize the central message of the Critik der reinen Vernunft, by the German Enlightenment philosopher Immanuel Kant (1724–1804) is this: there is no such road to metaphysical knowledge, for what rationalists suppose to be metaphysical first principles apply only within the realm of possible experience, and attempts to deduce from them conclusions about objects beyond that realm can therefore lead only to irresolvable contradiction.

Whether Kant's critique of the "mathematical" conception of the epistemology of metaphysics is defensible or not, a very simple observation raises a profound difficulty for that conception. It is the observation with which this section opened—namely, that metaphysicians have failed to reach consensus on any important question. Metaphysics thus stands in stark contrast to mathematics, for, with very few exceptions, mathematicians are in complete agreement about what has and what has not been proved in their discipline. If the arguments of metaphysicians indeed consisted in the deduction of conclusions from premises that were self-evidently true by principles that were self-evidently valid, metaphysicians should agree about metaphysics in a way comparable to the way in which mathematicians agree about mathematics.

Metaphysical propositions are known in the same way that propositions in the physical sciences are known. That is, metaphysicians collect data that are relevant to the questions that concern them and attempt to construct theories that account for those data; having constructed such theories, they accept those that best account for the data and which best satisfy various constraints on the notion of an "acceptable theory". The theories that are obtained by this method constitute metaphysical knowledge.

The foregoing conception of the basis of metaphysical knowledge has been, and continues to be, common among Anglophone metaphysicians of the late 20th and early 21st centuries. It seems, however, to founder on the same embarrassing observation that undermines the "mathematical" conception of metaphysical knowledge: metaphysicians have failed to reach consensus on any important question. If the "scientific" method of investigation were indeed a source of metaphysical knowledge, should one not expect a degree of convergence of belief among metaphysicians that approximated the convergence of belief among physical scientists? But there is no such convergence of belief among metaphysicians.

Metaphysical knowledge is knowledge of certain a priori structures that are somehow inherent in the human mind—or perhaps in the mind of any finite rational being. Such is the "transcendental" metaphysical knowledge that is promised by Kant's critical philosophy. Examples of such structures are the "forms of intuition"—space and time—and the "categories" of the pure understanding, such as causality and substance. In some respects, transcendental metaphysical knowledge would be a special case of the kind outlined in category 2 above. The data in question are provided by the existence of synthetic propositions that are known a priori. Among synthetic a priori propositions are the propositions of geometry, arithmetical propositions, and certain very general principles that are presupposed in the natural sciences, such as "Nothing comes into existence without a cause." The last proposition may serve as an example of what Kant sets out to explain. Kant argues that, although the concept of "having a cause" is not included in the concept "coming into existence," the proposition "Nothing comes into existence without a cause" is known a priori, because (a) it is known, and (b) it cannot be known a posteriori (on the basis of experience). Transcendental metaphysics is thus essentially an explanatory theory, one designed to explain how such "synthetic a priori knowledge" is possible. But transcendental metaphysics implies that transcendent metaphysics is impossible: in making explicit the a priori forms of intuition and the categories, it shows that they apply only to possible objects of experience.

The Kantian epistemology of metaphysics faces a difficulty that is very much like the difficulties that face the two epistemologies previously discussed: very few contemporary philosophers accept the theses that, for Kant, constitute the content of transcendental metaphysics. In addition, one may ask: If those theses were indeed known by Kant—or by anyone—why have they not been generally accepted by philosophers? After all, scientific knowledge—once some person acquires it—can be passed on to anyone else who is willing to attend to the data and the arguments that were the basis of the first person's knowledge.

Metaphysical knowledge is knowledge of certain logical consequences of the general knowledge that one brings to the study of metaphysics. That is because certain metaphysical propositions that have been the subject of dispute throughout the history of the field can be seen to be logical consequences of propositions that people know to be true on non-metaphysical (and nonphilosophical) grounds. The basis of metaphysical knowledge is thus twofold. It consists partly in whatever the basis of "general" knowledge is (the knowledge that one brings to philosophy) and partly in whatever the basis of knowledge of logic is (the knowledge of valid logical inference). Moreover, it would have been a part of the task of epistemology to give an account of those two kinds of knowledge in any case. Thus, on the present conception of metaphysical knowledge, metaphysics raises no epistemological problems peculiar to itself. The conception presupposes that all metaphysics is what the British philosopher Sir Peter Strawson called "descriptive metaphysics." Descriptive metaphysics does not aspire to correct—or even to add to—the stock of general knowledge or of scientific knowledge; it attempts only to reveal the implicit metaphysical content of such knowledge.

An example of descriptive metaphysics is the following argument, which purports to show that the existence of universals is a logical consequence of scientific knowledge, specifically knowledge of biology. (a) The proposition "Some species are cross-fertile" is one item in the body of biological knowledge. (b) The proposition "There are species" follows logically from that proposition, by the same rule of logic that licenses the inference from "Some philosophers read each other's books" to "There are philosophers." (c) Species, moreover, are among the standard examples of things that, if they exist, are universals. (d) Hence, the existence of universals is a logical consequence of biological knowledge—and, in becoming aware of that fact, one comes to know that there are universals.

The preceding account of the basis of metaphysical knowledge seems to be straightforward and unproblematic—although, and perhaps because, it severely limits the scope of metaphysical knowledge. Nevertheless, that seemingly straightforward and unproblematic epistemology of metaphysics faces the same embarrassing difficulty as its alternatives: the problem of apparently irresolvable metaphysical disagreement. It faces that problem because every observation or argument that has been propounded to show that some metaphysical proposition is a consequence of general or scientific knowledge has been disputed by metaphysicians of the highest reputation.

If an epistemology of metaphysics is a theory that explains the basis of metaphysical knowledge, the search for such a theory must assume that there is a significant body of metaphysical knowledge for it to explain. And yet those who believe that such a body of metaphysical knowledge exists have yet to provide a satisfactory reply to the following simple argument: (a) If a significant body of metaphysical knowledge existed, all or most reputable metaphysicians would accept the propositions it comprised. (b) But metaphysicians have failed to reach a consensus or near-consensus on any important question. (c) Therefore, no significant body of metaphysical knowledge exists.

7

VEDIC METAPHYSICS

7.1 INTRODUCTION

It is a common observation that all types of life have some energy right at the time of birth. While a human infant on birth is normally found, crying but many newborn babies of other animals are found even walking along with their mothers. Somewhat similar phenomenon is observed in the other life like, birds, fishes and sea animals.

Is it only the intake of the food by the mothers, the source of energy of all new born babies or is there some thing else which can be termed as the energy principle and operates through the divine Nature ? What had made Dante to say, "man is a divine animal" and Aristotle to say, "man is a social animal"?

If for Socrates, Plato and Dante the soul is the individual's real Self, why the same soul Plato's greatest and the most renowned disciple Aristotle could not find and in lieu found entelechy moving towards perfection both in animate and inanimate life?

What made Oscar Wilde to say, "I am the only person in the world I should like to know thoroughly" ? This "I" certainly cannot be the gross body. If due to any unfortunate major accident the eyes become blind, either the legs or hands are badly injured and finally amputated, this "I" still remains intact even though gross human body is not complete.

This "I" can at best be brain, mind, intellect, will, ego, spirit or soul. The Vedic metaphysics tells us that it is the soul which makes a person separate and distinct individual. The spirit is the cause of his/her social being and along with panch-mahabhuta - five main elements and three gunas of purity, activity and stupor is the cause of good or bad actions of the individuals.

Thus in the Vedas an individual is both divine and social animal, capable of doing both good and bad deeds. Metaphysicists like Sankracharya of India, Lao Tse of China and Immanuel Kant of Germany had found gross body as part of the phenomenal world of Maya or Phantasmagoria.

They found the real extended " self " of the individual in his / her divine soul. However, the Vedic metaphysics describes clearly all these concepts relating to soul (atma), manifested soul (jivatma), spirit (jiva), ego (ahamkar), intellect (buddhi), mind (manas), gross body (sthool sharira) etc.

According to Vedic metaphysics human gross body has an inward looking mind (manas) and an outward looking mind (etani), they both work together as engine of the gross body. Five Koshas are the wheels and the path to be covered is not known to the engine (mind).

This path is only known to the manifested soul (jivatma), which is nearest on self-realization and farthest when phenomenal and the material worlds become the closest. The various hymns in the Vedas as well as most of the 108 Upanishads and Bhagavad Gita explain the concept in details. The essence of the metaphysical concepts of soul as part omniscience principle and spirit as energy and life principle. Both are divine instruments in the human gross body along with ego, intellect and mind.

The various terms are used in the Vedas like Parmatma, Purushuttam and Brahma for Supreme Self - the formless and ineffable God. Jivatma, Purusha and Brahmansh are used for the real higher self in the human body, which is the manifested immaterial soul. For the soul still not manifested in the human body, instead of jivatma, the term atma is mentioned.

For the divine spirit in the gross body the epithets like Jiva, asu, kundalini, - manas, prana etc., have been given. This spirit is found in all parts of the subtle and gross body including nails and hair. Spirit is the one, which makes things move not only in animate but also inanimate life. Kundalini is described as the powerhouse of coiled energy. Asu and Pranas are ten different vital breaths of which, seven are subtle and three are gross.

While making a distinction, Vedas make it clear that God is both the Supreme Soul and Supreme Spirit. In the Vedic metaphysics this is described as Parmatma - the supreme soul and Mahadeva - the supreme spirit. Being subtle in nature, His spirit even pervades in ideas, thoughts, actions, assemblies etc.

Very often we say "the spirit behind" the action, thought, desire or religious or political assembly is XYZ... Thus the spirit of God pervades not only in animate or inanimate life but also in subtle concepts, ideas, actions etc., and makes the formless and ineffable God as omnipresent. The individuals who know and understand the spirit as the powerhouse of creativity in them create subtle waves of nobility, love, and harmony even if they sit silently in any gathering or assembly.

Patanjali- 's Yog shastra, Manu Smriti, - Puranas, Brahma sutras and other scriptures agree with Vedas that the Spirit of God pervades every where in all animate and inanimate life. However, right from Charvakas days, a large number of Hindus find difficult to accept any energy principle in the inanimate life. For them matter in any form is inert. The acceptance of God's spirit in the inanimate life is more as a faith but not conviction.

The Newtonian classical mechanics made it still more difficult for the masses to accept life in iron, silver or gold, granite and marble stones. Newton held atom as the building block, which could not be further divided into sub-atomic particles and confirmed that atom is inert. The classical mechanics of Newton and Tyndall encouraged the western philosophers and thinkers to decry idealism of Socrates, Plato, Immanuel Kant and many others in the East and West.

The worst controversy arose when the experimental biologist Charles Robert Darwin (1809-82) agreed that the evolution of the present day morphology had been built up by the gradual and opportunistic mechanism of natural selection that he described as "survival of the fittest". He also held this principle rules this gross universe and there is no such thing as subtle or subtlest of the subtle.

His dis¬covery of survival of the fittest in this universe has no relationship to soul, spirit or other metaphysical concepts contained in the Holy Scriptures and various philosophies, particularly those of pure and objective idealists. Vedic metaphysics that had reached its pinnacle of glory during Sankracharya, Ramanajum, Vallabh 's periods got a set back.

7.2 METAPHYSICS IN 'THE VEDAS'

Vedas etymologically translate to 'Knowledge or Wisdom'. These were written from 1700 BC to 500 BC in northern India and were completed in various stages. They are basically divided into 4 different books.

The Vedic corpus is basically divided into four parts. Samhitas (Mantras & Benedictions) are the oldest layer of the Vedas, Brahmanas (text pertaining to Rituals), Aranyakas (wilderness texts pertaining to ceremonies) & Upanishads.

Rig Veda is the oldest composed around 1500 BC and mainly deals with hymns for older Gods like Agni, Indra, and Vayu. However, under Nasadiya Sukta it asks some fundamental Metaphysical questions, like, What is the origin of the Universe? to even questions like, What is the virtue of charity in society? But my favorite which epitomizes the curious scientific thought of the philosophers writing these is, Did we create God or did God create us?

Saam Veda: Based on Rig Veda is divided into two parts, four melody portions in the first part and three verse portions in the second part.

Yajur Veda: These are archaic prose mantras independent of Rig Veda.

Atharva Veda: Atharvaveda has less connection with sacrifice. It has numerous verses on magic spells and charms, such as those to gain better health, finding a good spouse, peaceful social and family relationships, and others. The second part of the text contains speculative and philosophical hymns. The Atharva Veda has been the primary source for information about Vedic culture, the customs, and beliefs, the aspirations, and frustrations of everyday Vedic life, as well as those associated with kings and governance. The text also includes hymns dealing with the two major rituals of passage – marriage and cremation. Finally, the Atharva Veda dedicates a significant portion of the text asking the meaning of a ritual.

Out of about 17,000 hymns in the Vedas, almost half of them relate to Metaphysics. The remaining hymns pertain to mantras (in praise of God), scientific rituals/ceremonies and Aryankas (simple rules for the society). Other texts written around the same time in Indian sub-continent are equally flattering towards Metaphysics, which both Ramayana & Bhagavad-Gita call it as the supreme science.

Vedic Metaphysics wisdom mostly pertains to the cosmic workings. From microcosm to macrocosm, it tries to deal with the whole range. It comprehensively covers material, temporal, secular & spiritual. In the material knowledge there is a mention of global trade for the welfare of mankind and need to manufacture fast moving aircraft, vehicle, and ship by the industrialists with the help of scientists.

"Let noble thoughts come to us from all sides." R.V. 1-89-1

Vedic religion aims at creating scientific outlook in society over savage rituals. Myths and aimless rituals are not advised. Though primitive in its understanding of the cosmos the idea that basic aim of religion is to understand the working of the universe is fundamental to it.

Dharma is Rta- the cosmic laws of social, moral and physical order. Having created these laws, even formless and ineffable God follows these laws sternly.

The subtle divine Nature is our supreme Mother. She is described as Maha Maya- the creator of phantasmagoria. Under its vehement effect the human senses find this phenomenal world as real.

The main Vedic concepts like enlightened liberalism based on the philosophy of 'idd nan mmam'– nothing for self, all for society. Difference between soul and spirit, personal and impersonal God have also been mentioned in the Vedas.

Vedas covers the Cosmic working at all levels, from microcosm to macrocosm. It covers material, temporal, secular, spiritual, and divine knowledge by blending both the inner and outer worlds.

In the concept of Brahma as the only Reality, the highest metaphysical ethics is reached as all else is Maya. It treats Maya is a Creative Art. Isvaraya Upanishad explains this as the illusory, which appears real to human senses.

The Vedic metaphysics tries to answer doubts which most of us have. Why and how this world was created? How am I? Where did I come from and where shall I go? Why I am here? Do I come back in the next life or just become manure for the earth? What is an ideal way of life?

Vedas are exhausting in their treatment and for the breadth of material. The sages & philosophers, who did speculative transcendental research to find permanent truths, came to the conclusion that any effort to change or modify them will be futile. Such truths are a-priori. Only empirical knowledge based on sense experience like physical sciences gets modified with the passage of time. They tried to delink projecting our sense experience and perception, criteria and even preferences into an objective universe, which they thought lead to the concept of morality relative to each individual.

The main pillars of the Vedic philosophy are (a) 'Idd Nan Mmam', nothing for self, all for society, (b) 'Rta' which literally means Laws, are the cosmic laws of social, moral and physical order, (c) trinity of values – 'Dharma', 'Artha' & 'Kama'.

So, in the Vedic tradition, dharma is just not a theology, but covers morals of the individuals, ethics for society and creation of material wealth for the society.

Although Shrutis deal with a large number of subjects including education, physical sciences, medicine, military science, various duties of kings, queen commander in chief, bureaucrats etc., apart from many metaphysical concepts,

The real "self" of the individual described in Vedas as Jivatma has no sex. It is only the gross body, which is described, as "body self" is either male or female. While the subtlest than the subtle human manifested soul is imperishable, the gross body is perishable. All our good or bad actions, desires and thoughts create subtle atoms ('kanu') and particles ('tanmatras') which go on accumulating like, iron files on the magnet on the human soul and become the cause of rebirth till one achieves Moksha -the stage of eternal bliss.

During the later periods of ancient Indian history, there had been a large number of women who became 'Rshikas'. In the 8th century B.C Rshika Gargi was one of the most prominent female metaphysicists.

There is also some mention of distance between the Sun and the earth, moon and the earth, the movement of the earth around the stationary Sun, force of gravity, magnetism, electricity hidden in water, telegraphy and particle physics. There is life in the gross atoms and particles, as the spirit of God is present in all animate and inanimate life. There are a large number of hymns in the Vedas particularly in Rig-Veda relating to medicines, toxicology, physical sciences, agriculture science etc.

Another very important subject covered comprehensively in the Vedic metaphysics relates to 'Maya', which is the cosmic illusion of this material and phenomenal world. Most of the Upanishads, Bhagavad Gita and metaphysics of seers like Sankracharya, Ramanajum and a few schools of Indian Philosophy deal in great details about Vedic Maya. This is like Kant's Phantasmagoria. The ancient Greek objective idealist Plato also found the material world as a phenomenon and not absolutely real. The pure idealist in Kant described the world as phantasmagoria. There is only ONE ineffable and formless Supreme Reality, all else is Maya. The personal God 'Isvara' with a large number of attributes is the God of the phenomenal world of Maya, which leads to many gods ('Ishvar') all compatible with the Supreme Reality but lesser deities. 'Ishwar' is the god who rewards, punishes and arbiters on your past and present Karma for your successive births, which take place in the realm of Maya. The Vedic metaphysics advises to get your "self" released from 'Ishvaras' and get absorbed in 'Brahma' the formless Supreme Reality. These 'Ishvaras' help the individuals on the path of knowledge, in their movement toward One God.

Based on the degree of material, spiritual and divine knowledge one has acquired, the Vedic metaphysics is understood differently by different people. It is perhaps for this reason that six schools of Indian Philosophy explain the Vedic metaphysics differently.

There is absolute monism and pure idealism in the metaphysics of 'Advaita Vedantists'.

There is subjective idealism in the philosophy of qualified monists popularly known as 'Vashisht Advaita Vedantists'

Complete dualism in the 'Paramarth' of 'Dvaita Vedantists'.

There is atheism in the 'Sankhya' and 'Mimansa' darshana as these two schools of Indian philosophy tend to give more emphasis to divine Nature- the supreme. These two schools find God as "Adrsta"- the unseen Cosmic Force.

'Yog Shastra' of Patanjali tells us that Vedic God is only the first preceptor (Guru) who helps us in learning faster the yogic concept of 'Kaivalya' – knowing your self. On self- realisation at the stage of Kaivalya, a person can even communicate with his/her soul. Even without reaching this stage of kaivalya, many modern gurus, their disciples and followers claim that they see the inner light and can communicate with their souls while under the vehement effect of Maya. It is nothing but hallucination- a stage when a person gets into utter darkness as per 'Isa Upanishad'.

The Vedic metaphysics give great emphasis to 'Prakrti'– the divine Nature, which is the supreme mother of all the animate and inanimate life. The inanimate life is not inert except that it has not even one indriya (sense) like plants and trees, which are animate having one sense of touch.. Animals have five senses. While human beings have also five senses but they have in addition divine soul, which makes the man as "divine animal". However, the supreme mother Prakrti provides all the five- mahatatva (main elements) of water, air, fire, ether and earth to all animate and inanimate life along with the spirit, which is the energy principle of all kinds of life. It is this energy, which is subtle in form and stays in "Shuniya"- void of all particles ('tanmatra') and atoms both subtle and gross ('kanu and anu') and provides the properties and characteristics to all elements ('bhuta'). Vedas tell us that there are 720 elements in this gross Universe. Even the abstract metaphysical concepts relating to Time and Space (Kala and Dis), human life is too short and insignificant in time and space and need to get detached to matter by avoiding material excesses has also been brought out.

To appreciate the Vedic metaphysics understanding of the human body consisting of 23 objects/subjects as our outer gross and inner divine instruments is necessary. According to Bhagavad Gita, these are 5 senses, 5 sense organs, 5 maha bhutas (ether, air, fire, water and earth), three gunas (primordial matter) of purity, activity and stupor, mind (both inward and outward looking), buddhi (intellect), ahamkar (ego), jiva (spirit) and jivatma (manifested soul).

8

BUDDHIST METAPHYSICS

8.1 INTRODUCTION

Buddhist metaphysics and modern symbolic logic might seem strange bedfellows. Indeed they are. The thinkers who developed the systems of Buddhist metaphysics knew nothing of modern logic; and the logicians who developed the panoply of techniques which are modern logic knew nothing–for the most part–of Buddhism. Yet unexpected things happen in the evolution of thought, and connections between these two areas are now emerging.

At the time of the historical Buddha, Siddhartha Gautama, c. 5c BCE, a common assumption was that there are four possibilities concerning any claim: that it is true, that it is false, that it is both true and false, that it is neither true nor false. The principle was called the catuskoti.

We know this because, in some of the sūtras, the Buddha's disciples ask him difficult metaphysical questions, such as the status of an enlightened person after death: does the person exist, not exist, both, or neither? They clearly expect him to endorse just one of these possibilities.

Until recently, Western logicians have had a hard time making sense of this. Standard logic has assumed that there are only two possibilities, true (T) and false (F)–tertium non datur. In particular, there is nothing in the third koti (both true and false), and even if there were, it would be in the first two as well (true, and false). So the kotis would not be exclusive.

To those who know a little modern non-classical logic, however, the matter is easy. There is a system of logic called First Degree Entailment (FDE). This is a four-valued logic whose values are exactly those of the catuskoti (t, f, b, and n). Standardly, they are depicted in a diagram as follows:

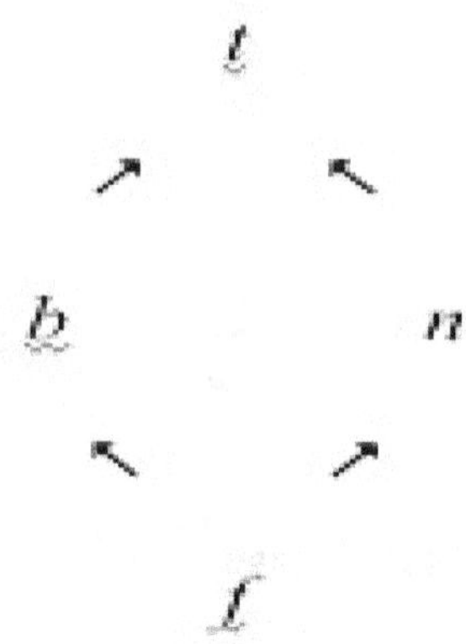

Fig.1 Modern Non-Classical Logic

Given this ordering, the values interact with other standard machinery of logic, such as negation, conjunction, and validity.

Matters, however, don't end there. The Buddha refused to answer such tricky questions. Some sūtras suggest that this was because it was simply a waste of time. Others hint at the possibility that sometimes none of the four kotis may be the case.

And this is just what later metaphysics suggested. Nagarjuna is perhaps the greatest Buddhist philosopher after the Buddha. His Mūlamadhyamakakarika (Fundamental Verses of the Middle Way) was a text which was to exert a profound influence on all subsequent Buddhisms. And in this, he appears to say that some things are simply ineffable. The state of an enlightened person after death, for example, is such.

This adds a fifth possibility, i, which is none of the above. It is distinguished from the first four, in that if something, A, has the value i, so does any more complex sentence formed out of A.

This still isn't an end to the matter, however. For Nagarjuna, and those who follow him, not only claim that some things are ineffable: they explain why they are so. Clearly, speaking of the ineffable is a most paradoxical state of affairs. It would seem to show that some claims can take more than one of the five values, such as both t and i. (And lest it be thought that this is simply a feature of Eastern mysticism, one should note how many of the great Western philosophers have found themselves in exactly the same situation: Aristotle (prime matter), Kant (noumena), Wittgenstein (form, in the Tractatus), Heidegger (being).)

Again, modern techniques of non-classical logic can show how to make sense of this possibility. In standard logics, things take exactly one of whatever values are on offer. But in a construction called plurivalent logic, things can take more than one such value. The plurality of values interact in a perfectly natural and sensible way with the other logical machinery.

What to make of these matters? Philosophers may certainly argue about this. However, one thing is ungainsayable: Buddhist metaphysics and formal logic can profitably inform each other.

FDE was a system of logic known independently of Buddhist considerations. But the development of the five-valued logic above, and plurivalent logics, were motivated, at least in part, by trying to make sense of the Buddhist picture. The metaphysics can therefore stimulate novel developments in logic—developments whose interest is not simply restricted to applications to Buddhist metaphysics.

Conversely, those with a suspicion of metaphysics in general and Eastern systems of metaphysics in particular, might be tempted to write off such enterprises as logically incoherent. They're not. The techniques of modern logic certainly don't show that these pictures of reality are true. However, they show that they're as logically coherent as can be – and, moreover, they allow us to articulate the views with a precision and rigor hitherto unobtainable, and hence to understand them and their consequences better.

Buddhist views are not, of course, the only views that can work in this dialectical way. But because Buddhist views are little known in Western philosophy, they provide a particularly fruitful domain of inquiry.

And what would the Ancient Buddhist metaphysicians themselves have made of such developments in logic? We'll never, of course, know; but my own guess is that they would have very much appreciated the enlightenment which such techniques can bring.

8.2 THE ANTI-METAPHYSICS ARGUMENT

Most people who argue that the Buddha wasn't interested in the nature of reality provide two examples from the Pali Canon.

In the Cula-Malunkyovada Sutta, a monk named Malunkyaputta declared that if the Buddha did not answer some questions -- Is the cosmos eternal? Does a Tathagata exist after death? -- he would give up being a monk. The Buddha replied that Malunkyaputta was like a man struck by a poisoned arrow, who would not have the arrow removed until someone told him the name of the man who had shot him, and whether he was tall or short, and where he lived, and what sort of feathers were used for the fletchings.

Being given answers to those questions would not be helpful, the Buddha said. "Because they are not connected with the goal, are not fundamental to the holy life. They do not lead to disenchantment, dispassion, cessation, calming, direct knowledge, self-awakening, Unbinding."

In several other places in the Pali texts, the Buddha discusses skillful and unskillful questions. For example, in the Sabbasava Sutta, he said that speculating about the future or the past, or wondering "Am I? Am I not? What am I? How am I? Where has this being come from? Where is it bound?" gives rise to a "wilderness of views" that do not help liberate one from dukkha.

8.3 THE PATH OF WISDOM

The Buddha taught that ignorance is the cause of hate and greed. Hate, greed, and ignorance are the three poisons from which all suffering comes. So while it is true that the Buddha taught how to be liberated from suffering, he also taught that insight into the nature of existence was part of the path to liberation.

In his teaching of the Four Noble Truths, the Buddha taught that the means to be released from suffering is practice of the Eightfold Path. The first section of the Eightfold Path deals with wisdom -- Right View and Right Intention.

"Wisdom" in this case means seeing things as they are. Most of the time, the Buddha taught, our perceptions are clouded by our opinions and biases and the way we're conditioned to understand reality by our cultures. Theravada scholar Wapola Rahula said in What the Buddha Taught that wisdom is "seeing a thing in its true nature, without name and label." Breaking through our delusional perceptions, seeing things as they are, is enlightenment, and this is the means of liberation from suffering.

So to say that the Buddha was only interested in releasing us from suffering, and not interested in the nature of reality, is a bit like saying a doctor is only interested in curing our disease and is not interested in medicine. Or, it's a bit like saying a mathematician is only interested in the answer and doesn't care about numbers.

In the Atthinukhopariyaayo Sutta, the Buddha said that the criterion for wisdom is not faith, rational speculation, views, or theories. The criterion is insight, free of delusion. In many other places, the Buddha also spoke about the nature of existence, and of reality, and how people could free themselves from delusion through practice of the Eightfold Path.

Rather than say the Buddha was "not interested" in the nature of reality, it seems more accurate to conclude that he discouraged people from speculating, forming opinions, or accepting doctrines based on blind faith. Rather, through practice of the Path, through concentration and ethical conduct, one directly perceives the nature of reality.

What about the poison arrow story? The monk demanded that the Buddha give him answers to his question, but receiving "the answer" isn't the same as perceiving the answer oneself. And believing in a doctrine explaining enlightenment is not the same thing as enlightenment.

Instead, the Buddha said, we should practice "disenchantment, dispassion, cessation, calming, direct knowledge, self-awakening, Unbinding." Merely believing in a doctrine is not the same thing as direct knowledge and self-awakening. What the Buddha discouraged in the Sabbasava Sutta and Cula-Malunkyovada Sutta was intellectual speculation and attachment to views, which get in the way of direct knowledge and self-awakening.

9

METAPHYSICS IN CHINESE PHILOSOPHY

9.1 INTRODUCTION

The Stanford Encyclopedia of Philosophy itself seems ambivalent. There is this entry, but instead of a corresponding entry for "Metaphysics in European Philosophy" there is an entry for "Metaphysics" that draws exclusively on European traditions. Rather than argue for the legitimacy of my topic in the abstract, I will explicate specific issues and positions from various Chinese philosophers, leaving it to readers to decide how well they fit the category of metaphysics. Nonetheless, I cannot entirely ignore the problem of applying the term "metaphysics" so far beyond its Greek origins, and so a few preliminary comments are necessary.

If we designate an area of philosophical inquiry concerned with the ultimate nature of reality—its being, origins, components, ways of changing, and so on—there is no question that Chinese philosophers addressed issues within this domain. The question is, do we apply the label of "metaphysics" to philosophical arguments within this domain, or do we reserve the label for some specifically European approach or theory. The latter faces an obvious problem—any definition broad enough to include all European approaches will include some Chinese theories, just as any definition narrow enough to exclude all Chinese approaches will exclude some European philosophers that everyone would agree addressed metaphysics. The choice of definitions is as much about rhetoric and power as the facts. If "metaphysics" labels the broader domain, then studying metaphysics involves engaging other cultures, and an SEP entry on "Metaphysics" that discussed only ideas derived from Europe would be biased and partial. In contrast, if "metaphysics" labels a specifically European view, then courses, books, and encyclopedia entries on metaphysics legitimately exclude other cultures. Aside from the political consequences of that exclusion, it obscures the many areas in which the traditions can be placed in fruitful dialogue. Those factors support using an expansive definition of metaphysics. While applying "metaphysics" to both European and Chinese thought is more inclusive, it risks obscuring the differences between the two. More insidiously, it erodes these differences by presenting Chinese philosophy in European terms. That is a legitimate worry, but I think the greater danger is in ignoring what Chinese Philosophy might contribute to discussions of metaphysics.

If we do take Chinese Philosophy as having metaphysics, we must strive to avoid distorting it to fit into European terms. One problem is in selecting topics. A simple approach would be to list the main topics in European metaphysics and then see what Chinese philosophers have to say about them. The results would be disappointing, and this approach would miss what is most interesting about a cross-cultural perspective, which is its ability to raise new questions. Within this entry, I have tried as much as possible to follow and explicate the main issues that arose when Chinese philosophers were concerned with the nature of reality. I have then made brief gestures toward how these issues might connect to metaphysical problems in the European tradition.

A second problem is with drawing boundaries. A label like "metaphysics" refers to certain human practices at the same time that it draws boundaries around those practices. While Chinese philosophers engaged in the kinds of practices that metaphysics refers to, they did not draw the same boundaries. Isolating metaphysical inquiry from practices of self-cultivation, for example, would have struck almost any Chinese philosopher as odd, if not dangerous. There is no native Chinese term marking the same boundaries as "metaphysics" in European philosophy. When Chinese encountered the term, it was translated by way of Japanese as xing er shang xue, literally, "the study of what is above forms." "What is above forms" had long been a central concept in Chinese philosophy, originating in a passage from the "Xici" commentary on the Yi Jing, the Classic of Changes, which says: "What is above forms refers to the way; what is below forms refers to utensils ". The distinction between what is above forms and what is below forms has a vague resemblance to the distinction between metaphysics and physics, but it is more specific. The contrast is between the formed and the formless. Both were taken to be immanent in the world. They are two aspects of the processes of nature and would not be taken as distinct areas of inquiry. The implications of this view will appear across a range of metaphysical issues.

9.2 PROTO-METAPHYSICAL BACKGROUND: THE MANDATE OF HEAVEN

As far as we know, explicit metaphysical discussions began in China in the mid to late 4th century BCE with the Laozi (Daodejing) and associated texts. Before that, the two dominant philosophical movements were the Mohists and the Confucians (Ru). Both focused on political and ethical issues and showed little direct concern with metaphysical questions, but their discussions of the divine set the context for the emergence of metaphysical debates.

The two most relevant concepts are tian (heaven) and ming (the command; fate). The idea of tianming (the "Mandate of Heaven") first came to prominence in rationalizing the conquest of the Shang dynasty by King Wen and King Wu, who founded the Zhou dynasty in the eleventh century BCE. Heaven is described in anthropomorphic terms as having awareness, preferences, and values. As a political doctrine, the claim is that heaven will support rulers who help the people and will bring disaster on rulers who do not. On this view, the world works on consistent patterns that encompass ethical and political concerns. Human beings determine their own success or failure based on these patterns rather than depending on divine whim. The emergence of this view is commonly seen as a decisive moment in the formation of Chinese philosophy.

While heaven is presented as a willful and anthropomorphic being in the early parts of the Shangshu it was not transcendent in the sense of being external to the system of nature. The term tian simultaneously refers to the sky, the orderly movement of the heavens, and something that covers all things equally. The classical Chinese word for the "world" or "realm" is literally what is "under-heaven", tianxia. Since the actions of heaven occur through the world, the two are often difficult to distinguish. The primary expression of heaven's will is through human actions—to lose heaven's mandate is to lose their support. The eventual transition in conceptions of heaven from a willful deity to the causal patterns of nature was possible because heaven was never separate from those patterns. Debates were about the nature of heaven rather than its existence, and heaven remained a central term for philosophers holding widely different viewpoints.

The belief that heaven supported good rulers and punished the bad came into question as the Zhou dynasty fell apart. Something like the classical problem of evil arose as centuries of civil war and disaster made it difficult to believe that the world was structured along ethical lines (Perkins 2014). Three responses to this crisis can be distinguished. One, exemplified by the Mohists, defended and theorized the earlier view that heaven rewarded those who are good, defined as those who care inclusively for other people. The second emphasized that good and bad events come without reason and without concern for justice. This view was associated with a new meaning for ming, taking it not as a command but as something more like blind fate. The third position argued for the regularity of natural patterns but took those patterns as amoral. On this view, human beings remain in control of their fate, but what brings success is not ethical action, at least in a conventional sense. Versions of this view appears in the Laozi and in theories arising from practical arts like medicine or military strategy. While expressed in different ways in different times, the belief that nature follows consistent patterns that can guide human action became a dominant view across Chinese philosophies, while the relationship between these patterns and humanistic values became one of the main points of dispute.

9.2.1 The Cosmogonic Turn

Sometime probably in the middle of the fourth century BCE, a radical shift in vocabulary, concerns, and visions of the human took place. This new position has long been known from the Laozi, but recent archaeological discoveries show that the Laozi was just one of a number of positions that together constitute what we might call "a cosmogonic turn." These texts are the first we know of to directly question how the diverse things of the world arise and take form. This concern seems to have been bound up with de-centering and de-privileging human beings. As the Zhuangzi puts it:

> In stating the number of things we say there are "ten thousand," and human beings are just one of them. In comparison with the ten-thousand things, [human beings] are not even like the tip of a hair to the body of a horse.

This break with anthropocentrism went along with a shift away from humanistic values like rightness or ritual propriety and toward concerns with maintaining life, reducing desires, and acting spontaneously.

These cosmogonies share the following four assumptions:

1. The diverse things of the world ultimately trace back to a single source.
2. The generation of things happens spontaneously, without design or purpose.
3. The ultimate is immanent in the world and can be accessed in some way.
4. Between the ultimate and the myriad concrete things, there are intermediary steps, particularly a role for polarities and cyclical patterns.

Almost every example of cosmogonic thinking in Chinese philosophy shares these characteristics, so I will discuss each in some detail.

9.2.2 Monism

All Chinese thinkers who discussed ultimate origins took that origin to be unique. The best known name for this source is dao, which means path, way, or guide. Another important name is taiji the "supreme polarity." The term taiji appears in the Yi Jing as the original unity from which yin and yang emerge. It remained an important term, particularly during the Confucian revival in the Song dynasty. Positing a single source had a decisive influence on Chinese thought, as it implies an underlying unity and connection that easily threatens differentiation and division. One of the most persistent metaphysical concerns is the ontological status of difference and individuation. This orientation is the opposite of that in philosophies based on dualisms or ontologically independent substances, views that were dominant through most of European intellectual history. We can say that European metaphysics has tended to focus on problems of reconciliation, while Chinese metaphysics has been more concerned with problems of distinction.

There is some ambiguity in saying that the ultimate origin is one. The Laozi says that "the one" generates two, which generates three and then the myriad things, but claims that the one itself is not ultimate. It is generated from dao. That things are born from being, but being is generated from no-being. This reflects one of the earliest metaphysical debates—is this unitary origin a thing? There seems to have been advocates for each side, but the view that came to dominate is given as a principle in the Zhuangzi: "what things things is not itself a thing".

The terms you and wu are among the most important metaphysical terms in the Chinese tradition. They are often translated as "being" and "non-being", but wu refers not to radical nothingness but to the lack of differentiated beings. Thus in the context of the Laozi, Hans-Georg Moeller (2007) translates the two terms as "presence" and "non-presence", Roger T. Ames and David L. Hall (2003) translate them as "determinate" and "indeterminate", and Brook Ziporyn (2014) as "being-there" and "not-being-there".

Taking the ultimate as no-thing places it at or beyond the limits of language. How to speak of the ultimate without making it into an object or thing became another persistent philosophical issue. The Zhuangzi points out that as soon as you label something, even as no-being (wu), it becomes a thing that needs its own explanation:

There is being, there is no-being, there is not yet beginning to be no-being, there is not yet beginning to be not yet beginning to be no-being.

This dialectic between being and no-being was later taken up in a different form through Buddhist debates about emptiness, and it can be considered one of the central metaphysical problems throughout the Chinese philosophical tradition.

9.3 SPONTANEOUS GENERATION

If we take no-being as indeterminacy, then the problem of a first cause is not getting many from one nor getting something from nothing but rather how differentiation emerges from the undifferentiated. The common explanation appeals to another key metaphysical concept—ziran The character zi is a reflexive pronoun, and ran means to be in a certain way. Thus ziran means to be so-of-oneself or to be "self-so". Ziran excludes appeals to purpose, deliberation or design, and the rise of the term paralleled a displacement of heaven by terms such as dao. Using ziran to explain how things arise may seem like an evasion, not much different from replying, "that's just how it is." If we are to avoid an infinite regress of causes, though, the only possibility is to stop at something that just is the way it is from its own being. The role of ziran (self-so) is similar to the role of causa sui (self-caused) in European philosophy. While classical European metaphysics attempts to isolate self-causality in a transcendent God, the Chinese took it as the very nature of existence. In this approach, there are similarities with Spinoza's identification of being and conatus (striving) and even with Nietzsche's "will to power".

Two important points follow. First, existence is seen not in terms of abstract being but rather as sheng life, growth, birth, vitality. The fundamental role of sheng appears explicitly in the "Xici" commentary on the Yi Jing, which says that the foundation of the Changes is shengsheng, "generating and generating", "growing and growing", "living and living", or "natura naturans". This phrase inspired the famous description of nature as shengsheng buxi generating, generating, never ceasing! It is sometimes said that Chinese philosophy lacks ontology because philosophers were never concerned with being as such. It is more accurate to say that Chinese philosophers took dynamic organization as implicit in the very nature of being, rather than positing an external source for motion and order. This means that ontology is also cosmology, even biology.

Second, if spontaneous generation is the very nature of being, then one can legitimately attribute ziran to the ultimate or to things themselves. This immanence contrasts with the common division between God as self-caused and everything else as caused by God, a view which ultimately sees being as divided into two fundamentally different kinds. In employing a univocal conception of being as sheng, Chinese philosophies did not segregate self-generation from the world.

9.3.1 Immanence

The third common point is that the ultimate is immanent in the world. Verbs in classical Chinese are not modified for tense, and this introduces a fundamental ambiguity into all of these cosmogonies—while they can be read as describing something that happened in the past, they can just as well describe an ongoing process in which the generative function is always present. In one passage, Zhuangzi is asked where dao is and he replies that there is no place from which dao is absent. Pushed to give an example, he says dao is in ants and crickets. When asked to go lower, Zhuangzi says dao is in weeds, broken tiles, and even in piss and dung. Similar statements would later be made about Buddha-nature, particularly in the tradition of Chan Buddhism. The immanence of the source is demonstrated most of all by the fact that it remains accessible to cultivated people. In the Laozi, dao is something one can use in the world:

> Dao is constantly without name. Although in its unhewn simplicity it is minute, heaven and earth do not dare subordinate it. If princes and kings can preserve it, the ten thousand things will make themselves their guests.

It is difficult to find the right language to describe the relationship between dao and human beings. The dao is not external, so it is not a matter of getting or reaching it, and it is not an object that could be grasped. Since the self-so spontaneity to which dao refers is always present, what is required is a negative process of removing obstacles. Ziran is what remains if we free ourselves from striving and conventional goals. Thus this same process is described as wuwei, which literally means "lacking action" but refers to giving up actions that are coercive, effortful, or forced. The Zhuangzi describes this process as the "fasting of the heart/mind", which allows us to rely directly on vital energy (qi) and respond spontaneously to whatever appears before us.

9.3.2 Polarity and Cycles

All of the cosmogonies posit stages between the ultimate and the concrete myriad things of the world. If the ultimate ground of things is immanent rather than teleological, then concrete things must be explained through a gradual process of spontaneous differentiation. Having a series of stages also allows for degrees of differentiation within a connected whole. That provides an explanation not just for the concrete myriad things, but also for nature as a system.

The most common stage involves interaction between two forces. These polar forces could be specified in many ways—heaven and earth, hot and cold, dry and moist—but the pair that came to dominate is yin and yang. Yang originally referred to the south side of a mountain, which received the sun, while yin referred to the north side. Ultimately, yang was associated with the masculine, the forceful, and the bright, while yin was associated with the feminine, the yielding, and the obscure. Creativity followed from the interaction of yin and yang, analogous to sexual reproduction.

All of the cosmogonies include cycles and processes of return. One prominent model was the four seasons. The change of seasons places cycles of growth and decay into a broader context of continuous vitality. The change of seasons itself, though, was seen as expressing a more fundamental cycle between poles such as yin and yang. Cyclical change could also be theorized through the progressions of generation or overcoming among the five phases wood, earth, fire, air, and metal. Another version of this cycling between poles was the claim that when processes reach an extreme, they reverse. Yet another manifestation is that things emerge from a common source and ultimately return to that source. In all of these cases, cyclicality explains the sustainability and predictability of natural patterns.

9.4 IMPARTIALITY AND DIFFERENTIATION

No pre-Buddhist Chinese philosophers claimed that the qualitatively differentiated world we experience is an illusion, but their monistic metaphysics privileged connectedness and unity. While patterns of differentiation may be objective, individuation (i.e., what counts as a thing) is provisional and contextual. It is always possible to view all things as forming one whole or one body, and this unity tends toward equalizing things. From our contextual point of view, one thing can be said to be better, bigger, or more beautiful than another, but from a broader perspective all things have the same status as parts of a single body. This could lead toward skepticism of absolute values (as in the Zhuangzi) or toward an imperative to care for all things. Hui Shi (c. 380–305 BCE) is reported to have said: "Care overflowingly for all the myriad things; heaven and earth form one body". This can be seen as a radicalization of the Mohist claim that heaven generates all human beings and thus cares for them all equally, a point rooted in the early Zhou view of heaven as protecting the people. A progression toward more and more radical impartiality is mapped out in a passage from the Lüshi chunqiu a text compiled around 239 BCE. The passage begins with a statement of nature's impartiality:

The world is not one person's world but the world's world. The harmony of yinyang does not grow just one type. Sweet dew and timely rain are not partial to one thing. The birth of the myriad peoples does not favor one person.

This is followed by story contrasting Kongzi and Lao Dan:

A person of Jing lost a bow and was not willing to search for it, saying, "A person of Jing lost it, a person of Jing will find it, so why search?" Kongzi heard this and said, "If you leave out 'Jing,' then it is acceptable". Lao Dan heard it and said, "If you leave out 'person,' then it is acceptable." Thus it was Lao Dan who reached the utmost impartiality.

The level of impartiality attributed to Laozi eliminates the possibility of loss, leading to equanimity.

As the reference to Kongzi suggests, this tendency toward inclusivity threatened the humanistic ethics of the Confucians. Their main response addressed a gap in the cosmogonies exemplified by the Laozi. These explained the dynamism inherent in particular things and the broader patterns of nature, but they did not address the differences between kinds of things—what makes human beings differ consistently from dogs? The concept that arose to fill this gap and to justify Confucian humanism is xing Xing is most often translated as the "nature" of a thing or kind of thing, but it refers specifically to the way a thing responds spontaneously to its environment. In human beings, xing manifests itself primarily as desires and emotions, broadly labeled as qing. One of the key questions debated by the Confucians was the degree to which ethical concerns like care, respect, and shame were embedded in these natural spontaneous affects.

Xing became a foundation for theories of motivation but its roots are in metaphysics. Xing derives from heaven and is closely connected to sheng, the term meaning to live or generate. Xing moves from the generic creativity or vitality of nature to the specific life processes of kinds of things. More concretely, xing was conceptualized as the dynamic flow of qi (vital energy). One recently discovered text from the late 4th century BCE brings these levels together:

Although all human beings have xing, the heart lacks a stable resolve. It awaits things and then stirs, awaits being pleased and then acts, awaits practice and then stabilizes. The vital energies of pleasure, anger, grief, and sadness are xing. Their appearing on the outside is because things stimulate them. Xing comes out from what is allotted (ming) and what is allotted comes down from heaven (tian).

The term qi, translated here as "vital energy", is already well known in English. It was originally taken as one of several kinds of stuff, connected with air and breath, but it eventually became the dominant label for the basic stuff of the world, used to explain all kinds of dynamic processes, from the formation of heaven and earth to the patterns of weather to the processes of the human heart. It was closely connected with life and the generative power of nature. In this passage, human affects are the movement of this qi when stimulated by events in the world. This is part of the Confucian response to the focus on wuwei and reducing desires in texts like the Laozi and Zhuangzi—affects like sorrow and care arise spontaneously, by ziran. They are as natural for human beings as it is for water to flow downward or for trees to grow toward the sun.

It may seem that this topic has drifted from metaphysics into ethics, but human actions are not different in kind from the movements of other things in the world, and human motivation expresses the tendency toward growth inherent in the very nature of existence. The use of xing and ziran in relation to motivation differs from the concerns around free will in almost every way, but both arise as ways of explaining how human choices relate to the forces driving change in the rest of the natural world, or even how human choices relate to the very nature of being. In this way, both issues unite metaphysics and ethics.

Beyond its role in explaining motivation, xing helps explain the organization of the world into individuals and kinds. In a series of passages arguing with a rival named Gaozi, Mengzi shows that whatever would explain the natures of things must have its own dynamism and directionality, and it must explain the specific differences between kinds of things. On the one side, xing differentiates things in terms of coherent patterns of force, providing a contextual and provisional basis for individuation. One can refer to the xing of a human being but also the xing of the human mouth, to the xing of a mountain ecosystem or the xing of a single tree. On the other side, xing was generally used as a species concept—things of the same kind have the same xing. Mengzi's ethical philosophy is based on his belief that human beings share certain ways of responding to the world, all having the same xing. The status of species, though, was a point of controversy, linked to the question of whether or not all human beings could be held to the same standards. A more radical line of thought took each individual as having its own unique xing, a view rooted in parts of the Zhuangzi and developed later by Guo Xiang.

9.4.1 Correlative Cosmology

Near the end of the Warring States Period, new assumptions about cosmology and metaphysics appeared that dominated the Han dynasty and profoundly influenced the development of Chinese thought. This new view has come to be known as "correlative cosmology," but it was not a single cosmology as much as several cosmological principles. Its core elements are various schema for sorting phenomena into kinds and a theory of causality based on stimulus and response. These elements appear together in a paradigmatic statement from the Lüshi chunqiu:

Things of the same kind summon each other, those with the same vital energy join together, and sounds that match resonate. Thus if you strum a gong note other gong will resonate; if you strum a jue note other jue will vibrate. Use a dragon to bring rain; use the form to move the shadow. The masses of people think that fortune and misfortune come from fate. How could they know from where they truly come!

The categories used could be more or less specific, so on one side might be "human being" or "animal" while on the most general side all things could be classified as either yin or yang. Another of the most common categories were wood, earth, fire, air, and metal, known as the five phases. Yet another set were based on the Yi Jing, using either the eight trigrams or sixty-four hexagrams. These systems of categorization were eventually integrated, so that categories from one could be translated into the others. Given the underlying ontology of change and process, categorization is not based on inherent qualities or essences but on typical ways of acting and reacting—does it tend to expand or contract, work gradually or swiftly, manifest itself obviously or subtly? Since these traits are relational, the same "thing" may not always be in the same category (it might act like wood in one context but metal in another), and because they are dynamic, the categories give immediate information on how things can be controlled, influenced, or disrupted. The application of the categories depends on context and the context depends on our particular purposes, but they are meant to express real properties of things.

Consider the use of the most general categories, yin and yang. Yang labels the tendency to expand and dominate; yin labels the tendency to draw things in by yielding. Anything can be put in one of these two categories, but yin and yang are not inherent properties. The same thing that might be active and dominating in one relationship might be softer and yielding in another (as is commonly the case in Chinese medicine). The function of the labels can be compared to the way we label cause and effect. We can designate a cause and an effect in any change, but being a cause is not an essential property. Everything is simultaneously the cause of many effects and the effect of many causes. As cause and effect illustrate, even a set of binary categories can be helpful in analyzing situations, and yinyang could be specified in various ways. In the Yi Jing, lines representing either yin (a divided line) or yang (a straight line) can be combined into groups of three to form eight trigrams, or groups of six to form 64 categories. Each of these lines could be taken as more or less stable, leading to 4096 possible situations.

The conception of causality at work here has come to be labeled with the Chinese phrase ganying "stimulus and response". This way of approaching causality reflects the fact that existence is inherently active and dynamic: an effect is never purely passive or inert, and a cause works with the tendencies of the thing affected. The effects of a stimulus depend on the receptive and responsive capacities of the thing stimulated. One common model for ganying was resonance, as in the quotation above where the vibrations of one string stimulate vibrations in strings tuned to the same note.

The final element is the role of correlations. To place phenomena in the same category is to situate them as having similar functions in analogous configurations. To be yin is to have a relationship to something that is yang, to be wood-like is to stand in certain relationships to fire, metal, earth, and air. This way of categorizing allows correlations across what would seem to be very different kinds of things. For example, an illness that is yang (expansive or overactive) can be treated with foods that are yin (receptive or calming). This kind of thinking is rooted in a concern for acting in harmony with natural patterns, going back to the concept of shi, which refers to the seasons, to the temporal configuration of a given moment, and to the ability to act according to the demands of that moment. These correlations were based on empirical observations, but as the categories were integrated and extended to all phenomena, the connections become less and less apparent, sometimes seeming forced and arbitrary. For example, for the first month that starts spring, the Lüshi chunqiu correlates certain days of the month, scaly animals, the musical note jue, sour tastes, and the number eight, but it also recommends surveying the land to set the boundaries of fields and forbids conscripting groups for war or major construction. These are essentially rules for sustainability.

The model that dominated modern European thinking about causality—linear causality through collision (on the model of billiard balls)—was not central to Chinese reflections on causality (as it was not central in Europe before the late Renaissance). For Chinese philosophers, the paradigms for causality were things like the effects of music over a distance, the relationship between spring and the growth of plants, and the influence between a teacher and student. This orientation followed from belief that all things are interconnected and are ultimately composed of the same stuff, qi. It also reflected practical concerns—How does culture work so that people can live together harmoniously? How do we relate to nature in a way that is sustainable? Approaching causality from this direction, though, is notoriously difficult. Han dynasty philosophers were basically starting from what we might now call ecological thinking or theories of complexity.

Correlative cosmology posited a tightly ordered universe whose patterns could be grasped and mastered. The earlier quote from the Lüshi chunqiu ends with a powerful message: people think some things happen without reason or cause, attributing it to fate, but they are wrong. Nothing happens without a cause, and the system of causes can be known and controlled. This view placed human beings in control of their fates.

The elevation of human power appears in the way the system of correlations provided a metaphysical foundation for what would seem to be human constructs. One early example was the correlation of leniency and violence (or the civil, wen and the martial, wu) with the spring and the fall. This correlation made the use of violence a necessary and natural principle, while also restricting it to certain times. The paradigm for justifying social order through correlations is the Chunqiu fanlu, traditionally attributed to Dong Zhongshu (179–104 BCE). In this text, yin and yang remain complementary but shift from equal forces driving generation to markers for hierarchical positions in a system of correlated arrangements (Wang 2005). To give just one example:

The righteous between ruler and minister, husband and wife, and father and son all derive from the Way of yin and yang. The ruler is yang; the minister is yin. The father is yang; the son is yin. The husband is yang; the wife is yin. There are no places where the Way of yin circulates alone. At the beginning of the yearly cycle, yin is not permitted to arise by itself. Likewise, at the end [of the yearly cycle], yin is not permitted to share in [the glories] of yang's achievements. Such is the righteous principle of "joining." Thus, the minister joins his achievements to the lord; the son joins his achievements to the father; the wife joins her achievements to the husband; yin joins its achievements to yang; and Earth joins its achievements to Heaven.

This cosmology ends up doing much of the work that was done by anthropomorphism in the European tradition. In both cases, human culture is seen as mirroring structures at the foundation of the natural world. In the case of Europe, that foundation is seen as human-like in that we are made in the image of God, thus anthropomorphizing nature. In Chinese correlative cosmology, the opposite occurs, where specifically human phenomena are theorized as natural. In both cases, social and political hierarchies are given a metaphysical basis.

Han dynasty philosophy could still be quite critical. The elevation of yang over yin in the Chunqiu fanlu is partly driven by a desire to minimize state violence, which it correlates with yin. A more radical political alternative appears in the Huainanzi a text compiled around 139 BCE. The Huainanzi emphasizes the inherent tendency of being toward diversification. Correlative influences generate a world that is too complex to grasp or master; politically, that leads toward an argument for decentralization, minimal state power, and the value of diversity. Later in the Han dynasty, Wang Chong (27–100 CE) offers more fundamental criticisms, emphasizing the limits of human power to control and predict natural phenomena, the disjunction between virtue and worldly success, and the arbitrariness of explanations in terms of categories like the five phases. This argument is grounded in metaphysics. Wang Chong explicitly argues against beliefs that the world is ordered according to anything like a human design, claiming instead that everything arises spontaneously, by ziran.

The Han dynasty collapsed in 220 CE, leading to a long period of fragmentation, instability, and uncertainty. The dominant philosophical movement is known as Xuanxue, "Profound Learning." The term xuan means dark, obscure, or profound, but it also has a sense of what precedes any division, as it is used in the first chapter of the Laozi. The best known works from this movement are the commentaries by Wang Bi (226–249) and by Guo Xiang. Because of the centrality of the Laozi and Zhuangzi, the movement is sometimes known in English as "Neo-Daoism." Different philosophers held different positions, but the core metaphysical issue was how to understand dao as ultimate ground, particularly how to interpret descriptions of dao as no-being (wu) and how to understand the relations between dao and the concrete world of experience.

Xuanxue was important for establishing the metaphysical vocabulary used in later Chinese philosophy. One of the most important terms is li, which in its original use was a verb for laying out borders according to the contours of the land, or for carving jade according to its own patterns. Both Wang Bi and Guo Xiang use li as a technical term—for Wang Bi li refers to the patterns of coherence represented by the hexagrams of the Yi Jing, while for Guo Xiang li refers to patterns of differentiation that spontaneously arise in the world. Li was later used by Chinese Buddhists to refer to emptiness and by Neo-Confucians to refer to patterns of coherence.

A second key concept to arise is the pairing of ti and yong. Yong means use or function. Ti originally refers to an organized form, a pattern that can be recognized, or to a body or part of the body. The pairing of tiyong was applied in different ways, but what is most consistent through these uses is that ti is singular and yong is multiple. For example, the same piece of wood (ti) could have many different uses (yong): to fuel a fire, to build a house, to carve into bowls, and so on. As in this example, the multiplicity of yong comes through involvement in concrete circumstances or purposes. In this way, ti is usually less determinate than yong. In fact, the ultimate ti was generally taken as fully indeterminate, thus allowing for infinite determinate uses (yong). The contrast between ti and yong sometimes looks like a contrast between the thing itself and the various ways it can be used. Such a contrast is highlighted in the translation of ti as substance, but this translation is misleading in several ways. First, like yin and yang, the labels ti and yong are contextual and thus the very same thing might be considered as ti in one context but as yong in another. Second, the ultimate ti is almost never taken to be individuated—it is emptiness or vital energy or the patterns of coherence (li) that all things share. So, individuation occurs more on the level of yong than ti. It is worth noting that the phrase chosen to translate the ontos in ontology was benti literally the "root" ti, a term that was prominent in Neo-Confucianism.

9.4.2 Buddhist Metaphysics in China

A radical transformation in metaphysical views in China followed the introduction and incorporation of Buddhist philosophy, a process that began in the 1st century CE. On the surface, one might think that Buddhism opposes metaphysical speculation. In the famous parable of the arrow, the Buddha compares a student insistent on metaphysical speculation to someone who has been shot with a poisoned arrow but refuses to be treated until he knows who shot the arrow, why they shot it, where the arrow came from, and so on. The point is that we know that the central problem in life is suffering and we know that the cure is the elimination of self-centered desires. Nonetheless, Buddhists commonly claim that one can be released from such desires by seeing reality as it truly is. The truth to be realized is that there is no self as an independent and lasting being. Much of Buddhist philosophy can be read as a sustained attack on any kind of substance-based metaphysics. The existence of substances (and thus of the self as a real and lasting thing) requires three things: ontological separation between things (making the self distinct from others), internal unity and sameness over time (so it remains the same self). Buddhists attack all three of these, arguing that things are interconnected, lack intrinsic unity, and change endlessly. Their arguments invoke two of the main principles of Buddhist metaphysics—impermanence and dependent co-arising.

The process oriented metaphysics of Buddhism fit with the Chinese philosophical tradition, but it brought a level of precision and complexity honed through a long tradition of intense disputation and dialectic. Positions that had been taken for granted in China were articulated in detail and defended against alternatives that had never been a concern (such as the idea of an eternal and unchanging soul). New possibilities were introduced, including the claim that only mind is real. Of course, as Buddhism impacted Chinese philosophy, China transformed Buddhism, leading to schools of Buddhism that never existed in India. Buddhist metaphysics in China is complex and diverse and I will just focus a few examples.

Much of Buddhist metaphysics involves negotiating a middle ground between reification and nihilism. The problem appears in relation to the self—it may be true that there is no self, but surely there is something which grounds or generates or is the illusion of self. One early school (Abhidharma) argued the self is a label for what is really an aggregate of momentary elements and factors, known as dharmas (fa). Apparent wholes like the self can be mereologically reduced to constituent factors, just as a chariot can be reduced to its parts. A more radical view extended the critique of the self to any entity that might be taken as independently real or self-defined, in Buddhist terms, anything that might have svabhava, literally "self-being" or "self-nature". This denial of self-being follows from dependent co-arising, which claims that any event depends on and is bound up with others. Lacking an independent essence or ground, all phenomena are said to be empty, śūnyata or kong. Emptiness is meant to be a middle ground between affirming or denying the existence of things, but such a middle ground is difficult to articulate, and the history of Buddhism can be seen as a dialect between those proposing some kind of reality (accused of reification) and those rejecting it (accused of nihilism).

This dialectic is portrayed differently by different thinkers, depending on what they take to be the final position that encompasses all others. As an example, we can consider the progression given by Zongmi (780–841), a philosopher representative of the Huayan school but with close links to Chan. In his Inquiry into the Origins of Humanity (Yuanren lun) Zongmi begins with the view that each person has a soul that is reincarnated according to his or her actions. He critiques this position by analyzing this self in terms of its parts—the self cannot be identified with the totality of parts (since some parts are lost at death) nor with one part among many (which would make the other parts irrelevant and thus not really even parts). This leads into the next position, that bodily form and cognitive mind, because of the force of causes and conditions, arise and perish from moment to moment, continuing in a series without cease, like the trickling of water or the flame of a lamp.

According to Zongmi, this denial of the self cannot account for continuity, and in particular, the links between actions and consequences (karma). Some medium must hold the various elements together. That leads into the third position, that all the various changes are appearances of one mind, which is the ultimate reality. This was the position of the Yogacara or Consciousness-only.

The claim that mind is the ultimate reality tends back toward the side of reification and so the next step negates it. Zongmi's argument is typical of the Madhyamaka school (commonly known in Chinese as the "Three Teachings," Sanlun). If we identify this one mind with actual thoughts, then either both are illusions or both are real, but if we separate this mind from actual thoughts then we end up with thoughts that exist on their own and a mind that has no qualities whatsoever.

While the target here is mind as the ultimate reality, a similar argument can be used to attack anything claimed to be independent and unchanging. For Zongmi, this position denies any form of being and so it cannot end the dialectic. He calls the fifth and final position "The Teaching that Reveals the Nature," reinterpreting a term we have already seen, xing which here stands for the Buddha-nature This ultimate reality could also be referred to as the true mind or as the Tathagatagarbha, which literally means the "Womb of the Thus-Come" ("Thus-Come" being a common name for the Buddha).

While the last three positions all had prominent advocates in China, the final position became dominant and was shared by Tiantai Huayan, and Chan Buddhists. One obvious question is how this final position differs from the position of Mind-Only. Zongmi extends a line of argument common in Chinese Buddhism, which is to critique any form of dualism. Since nothing we experience is truly independent, simple, or unchanging, any metaphysics that posits substances with self-being requires a bifurcation between reality and appearances, or more specifically, between the substance itself and the various qualities or modes by which it appears. Buddhists argue against the coherence of any such division. This is the meaning of the common saying: "form is emptiness and emptiness is form." Zongmi pushes this argument further to argue against any split between reality and illusion or between enlightenment and non-enlightenment. This leads to the claim that all sentient beings are already enlightened—what is needed is not to change reality or gain anything but just to realize that we are already where we need to be.

The relationship between reality and illusion is one of the central debates within Chinese Buddhism. The most famous expression of this debate appears in the Platform Sutra of the Sixth Patriarch, a foundational text for Chan Buddhism. The star disciple first composes this poem:

The body is the tree of insight;

The mind is like a clear mirror.

Always clean and polish it;

Never allow dirt or dust!

Hui Neng (638–713 CE), the central figure of the text, writes this poem in reply:

Insight originally has no tree;

The bright mirror has no stand.

Buddha-nature is always pure and clean;

How could there ever be dirt or dust?

The contrast between the two poems marks the split between the gradual and sudden enlightenment schools of Chan.

How to discuss the relationship between the realm of experience and emptiness or Buddha-nature was another common theme in Buddhist philosophy. One of the more thorough and influential discussions is the Meditative Approaches to the Huayan Dharmadhatu. According to Dushun, the world of experience can be viewed in four different ways. The first is as phenomena or events (shi), which is equivalent to the realm of form. The second is as emptiness, referred to here with the term li. The point is that, in one sense reality can be seen as the multitude of phenomena but in another sense it can all be seen as empty, which entails a kind of equality and interchangeability. The third perspective addresses the relationship between emptiness and phenomena, using the relationship between water and waves as a metaphor. The two are mutually encompassing and mutually dependent, just as there is no water without waves and waves are nothing other than water, yet we can focus solely on the wave or solely on the water. The point is that designating a wave is not the same as designating water, yet these designations do not "obstruct" each other. The very same reality can be taken as the phenomena of everyday experience and as emptiness.

The fourth perspective brings us to a metaphysical issue that has recurred across the Chinese philosophical tradition—the interconnection of things. It is not just that emptiness and phenomena are mutually penetrating but that any phenomenon is penetrated by all others. The claim that any one thing includes all others is clearest on the level of causality and intelligibility. It might be for help in a course, or because you followed a link out of curiosity, or from a desire to better understand the context of the Laozi. But we could say the cause was the story of how your parents met—had that not happened you would not be reading this article. Or it might be the story of how my parents met, or the creation of the internet, or the founding of Stanford University, or the gravitational pull of the earth. If everything is interconnected, then anything could be given as a cause for your reading this article. What makes one answer better than another is determined only by the interests of the questioner. This shows how any one event implicates and arises with all others. Consider further, though, that all things are empty. They have no independent self-nature, so what it is to be that thing is explained entirely by all of the factors that allow it to appear as what it is. Thus it is not just that things depend on each other but that things include each other. This inclusion applies not just between any two events but also between any event and the totality of other events; the story of the whole universe can be explicated from any one point. Brook Ziporyn (2000) calls this "omnicentric holism"and it has similarities with Leibniz's claim that every substance (monad) implicitly expresses all others. This conclusion serves the soteriological purposes of Buddhist philosophy—if any one thing implies all others, then it is impossible to grasp only one thing. Grasping and attachment become incoherent. It is worth comparing this result with the elimination of loss through radical impartiality.

Coherence and Vital Energy in Neo-Confucianism

The interpenetration of emptiness and phenomena is an affirmation of the changing world in which we live. Thus, it is false to see Chinese Buddhism as life-negating or as denying the diversity of the world. Nonetheless, the metaphysics of emptiness is directed toward overcoming attachments. Diversity remains, but there are no individuals that could be held onto. The Confucian response, which became a dominant force in the Song dynasty (960–1279), was driven primarily by an aversion to these consequences. This reaction can be seen in three concrete positions—the acceptance of suffering and death as unavoidable, the differentiation of roles and norms within society, and the embrace of negative affects such as sorrow at the death of a parent or anxious concern for a child in danger. This Confucian movement is known in Chinese as the "Learning of the Way" and in English as "Neo-Confucianism."

While based on practical concerns, this Confucian revival was supported by metaphysical claims. In terms of the dialectal movement we have seen, the Neo-Confucians take all of the Buddhist positions as nihilistic. The first influential Confucian responses appear in cosmological arguments. We can take Zhang Zai (1020–1077) as an example. Zhang Zai's basic move is to argue that the ultimate, labeled as "Supreme Emptiness", cannot be nothingness but must be qi, and the fundamental characteristic of this qi must be a dynamic interplay between opposites. Zhang explains:

> The Supreme Emptiness of necessity consists of vital energy. Vital energy of necessity integrates to become the myriad things. Things of necessity disintegrate and return to the Supreme Emptiness. Appearance and disappearance following this cycle are a matter of necessity.

There are several important points to note in Zhang Zai's position. First, he explicitly argues that the only way to explain the origins of the world we experience is if dynamism and differentiation are the very nature of existence. For this reason, Brook Ziporyn argues that the ultimate for Zhang Zai is not qi but harmony itself. Second, by positing qi as fundamental, Zhang Zai shifts from questions about nihilism and reification back to the relationship between the formed and the formless. In making the formed and formless two modalities of existence, Zhang Zai allows that both are equally real. In this way, he reaffirms the importance of individuated things like parents and children.

Zhang Zai's cosmology echoes the earlier Chinese cosmologies discussed in Section 3. To say that the nature of qi is active differentiation is to say that the nature of qi is sheng, vitality or generation. This emphasis on being as a force of growth and vitality was a common point among Neo-Confucians, linked directly to the virtue of ren, humaneness or benevolence. To be humane is to support and extend the generative process of nature itself. Furthermore, there is a unity to things in the world, grounded in the fact that they are all made up of qi and they unfold in interlocking patterns of influence. As Zhang Zai puts it in the famous Western Inscription:

Heaven is my father and earth is my mother, and even such a small creature as I finds an intimate place in their midst. Therefore that which fills the universe I regard as my body and that which directs the universe I consider as my nature All people are my brothers and sisters, and all things are my companions.

Cheng Hao (1032–1085) later compares one who does not care for other things to someone who has lost sensation in their own limbs. Wang Yangming (1472–1529) extends these feelings of concern even to trampled grass and broken tiles.

Deriving inclusive care from the vitality and unity of nature appeared in the Warring States Period, but in opposition to Confucianism. For the Confucians, this focus on being as vitality needs an inherent order, which came through a reinterpretation of the term li The complexity of the term appears in the range of common translations: principle, patterns, coherence. Li often refers to something we should follow and in this sense it might be taken as principles, but li also refers to actual patterns of differentiation, not just to ideals. Considering that li is contextual and involves human purposes and perspectives, translating the term as "coherence" or "coherent patterns" probably best brings these various aspects together. Stephen C. Angle gives an excellent brief definition of li as "the valuable, intelligible way that things fit together". There are patterns of coherence in world, and these patterns define individual things, constitute nature as a system, and structure human society. Li describes the way that these patterns can be optimally harmonized or made to cohere so as to foster the human good.

The Neo-Confucian interpretation of li is a return to earlier Chinese views that took nature as having stable patterns that we can recognize and follow. The Neo-Confucians defend these earlier views against what they saw as Buddhist attacks by claiming that the differentiated structure of the world we live in is real. These patterns of coherence give specific form to our concerns and efforts, so that one can justify caring primarily for one's own parents while still taking all things as extensions of one's own body. Even so, the Neo-Confucian conception of li retains several Buddhist aspects.

The most obvious example is the claim that the totality of li is contained in any one thing. Cheng Yi (1033–1107) coined what became a standard motto: "Li is one but distinguished as many". This unity within diversity was illustrated with a metaphor taken from Buddhism—the moon reflects on many different surfaces, but it remains the one same moon. Since li refers to patterns of coherence and all things ultimately form one body, it follows that all things are mutually implicated, a point already discussed in relation to Huayan Buddhism above. The Neo-Confucians do not go so far as to say that each thing contains all other things, though. Any instance of coherence implicates all others, but events still have their own reality.

Another common Neo-Confucian claim with Buddhist overtones follows: if the totality of li is included within any particular thing, then it must exist within each person's heart/mind. One of the main disputes among the Neo-Confucians was on the precise meaning and significance of this point. Zhu Xi (1130–1200) argued that li is the nature found in the heart, while Lu Xiangshan (1139–1192) and Wang Yangming took li and the heart to be identical. This leads to a difference in emphasis between the two main schools of Neo-Confucianism regarding their recommended process of accessing li: the "School of Li" placed more emphasis on study and learning, while the "School of Heart/Mind" concentrated more on self-reflection. This difference, though, is one of emphasis: since they agreed that li could be accessed through our own heart and through things in the world, all Neo-Confucians promoted both self-reflection and learning.

Neo-Confucians always discuss li in relation to qi, vital energy. If li refers to patterns of coherence, qi is the stuff in which those patterns inhere. This distinction has functional similarities with Aristotle's distinction between form and matter, but the force of activity and change is with qi rather than li. That means that li must be immanent in qi in some sense. The precise ontological status of li and qi became one of the main metaphysical disputes among Confucian philosophers. One could easily claim that the distinction between patterns of coherence and the stuff that follows the patterns is conceptual rather than ontological. That is close to the view seen in Zhang Zai, for whom qi is active and inherently patterned. For Zhu Xi and Wang Yangming, though, li has ontological priority over qi. Zhu Xi writes:

> Fundamentally, li and vital energy cannot be spoken of as prior or posterior. But if we must trace their origin, we are obliged to say that li is prior. However, li is not a separate entity. It exists right in vital energy. Without vital energy, li would have nothing to adhere to.

The question of the relationship between qi and li extended into cosmogonies as well. While Zhang Zai took qi to be the ultimate, Zhu Xi claimed it was li. Since li is the locus of intelligibility and values, the question ultimately was whether causal and explanatory priority was with this coherence or if coherence was instead a product of the generative forces of nature (qi).

The distinction between li and qi provided a way to deal with the tension between the claim that the nature of human beings is good and the recognition that people need extensive self-cultivation in order to actually be good. Li is inherently good and is the same in all things. In that sense, we already have all that we need. The quality of qi, though, varies. Turbid qi—which manifests itself as selfishness and partiality—obscures li. Thus from the perspective of li human nature is good, but from the perspective of li and qi together, a person can be good, bad, or in between. On the psychological level, the work of self-cultivation is cultivating tranquility and impartiality; on the metaphysical level, one purifies qi so as to more perfectly express and participate in li.

This view of self-cultivation inherits the Buddhist claim that, since everyone possesses Buddha-nature, everyone is already enlightened, but merges it with Mengzi's claim that human nature is good. Unlike Mengzi, the process of self-cultivation is not a process of extension but rather of removing the obstacles that keep us from being what we already are. Phillip J. Ivanhoe has nicely captured this point by contrasting the views of self-cultivation for Mengzi and Wang Yangming as the difference between a model of development and a model of discovery. As in Buddhism, what needs to be removed are selfish and biased desires. Unlike Buddhism, and in line with Mengzi, this enlightened state does not lead to a calm state free of desires and negative emotions. Instead, the li within us is expressed as spontaneous and appropriate desires and emotions, including negative emotions like grief for a dying parent or concern for a child in danger.

9.4.3 Conclusion: Beyond "Chinese Philosophy"

If one turns to metaphysics in the twentieth century and beyond, it becomes necessary to distinguish "Philosophy in China" from "Chinese Philosophy". Like almost everywhere else in the world, the twentieth century in China was characterized by the incorporation of ideas from other cultures, most obviously from Europe. The terms "philosophy" and "metaphysics" were introduced as distinct concepts through translation and thus they become the objects of conscious reflection. Even scholars dedicated to "Chinese Philosophy" set themselves in dialogue with the West, so that almost all philosophy was more or less fusion philosophy. The main influences from Europe were Kant, Hegel, and Marx, but a wide range of philosophers were incorporated, including Nietzsche, Bergson, Dewey, and Heidegger. Almost all of the major philosophers in China—from Xiong Shili (1885–1968) to Feng Youlan (1895–1990) to Mou Zongsan (1909–1995)—were concerned with establishing a metaphysical foundation for Chinese Philosophy. These philosophers, drawing on resources from multiple cultures and traditions, are better suited for an entry on "Metaphysics" than one on "Metaphysics in Chinese Philosophy".

9.5 ONTOLOGY

9.5.1 Ontology is the Study of Being and Anything that Exists is a Being

Keeping it super simple, let's use the example of the table in your kitchen. It's all about the table as a being. Ontology is all about studying the table and everything about that table – in all scenarios.

Questions that would relate to ontology would be something like these:

- ⦿ What the table is?
- ⦿ How the table is?
- ⦿ How much the table is?
- ⦿ Where the table is (relating to other 'beings")

We could go even further and talk about the essential features of the table. As opposed to the accidental attributes of the table.

- ⦿ The table is solid, durable, straight. has 3-4 legs (essential features). It's square, 3 feet high and brown (accidental features)
- ⦿ Are there any fundamental entities of the table?
- ⦿ Yes, the table must have a straight level edge so things don't slip off of it. One could say a table would have to some form of legs or pillars to hold it up.

I'm over simplifying this by using a table as an example, but you get the point.

9.5.2 Cosmology

Cosmology is the study of the creation and development of the universe and humanity's place in it.

Humanity's place in the universe is the key question I had many years ago. The why was I here, what was my purpose – even more what is humanity's purpose on this earth – in this universe. These questions gnawed at my being. Which only led me to dig deeper.

Some people confuse cosmology and astronomy. Cosmology talks about the universe as a whole. This includes origin, evolution and the ultimate fate of the universe. Where as astronomy talks about celestial objects and phenomena. Which includes stars, galaxies gamma ray bursts etc.

Things like the Big Bang is an example of cosmology and one of the many theories of how the universe started.

Identity

This topic can get pretty deep as there are different terms for a variety of different types of identity. But if we keep it at the at the philosophical level we are looking at the relation each thing bears only to itself.

Let's continue with the table example we used above. Now, imagine we had two of the same tables. When looking at the identity of the table itself – a metaphysician might ask something like this:

This whole question brings up the topic of uniqueness and the fact that we all have our own identity. Of course with humans, we add in perspective and experience.

Look at identical twins – two human beings who share the identical DNA. They are even labeled "identical" and sometimes you can't tell them apart looking at them. But, they are not identical once you add in their experience and perspective with life.

Free Will

This is the term most of us will know. It is the ability to act at one own's discretion. The ability to choose, think and act voluntarily. To have free will is to believe that we as human beings are the authors of our own actions.

What an amazing outlook in life that is full of hope. The idea of free will is that the choices we make lead to opportunities and possibilities.

We have the ability to take action and determine the future we desire.

9.5.3 Mind and Matter

An easy way to think about this topic is to recall the common saying "mind over matter". The mind is a powerful thing, and metaphysics looks at how the mind relates and affects matter.

Years ago I worked for this company, where the owner would tell us "you are what you think about all day long". This was his mantra and he would tell us this all the time. If we thought about negative things in minds, then negative things would show up in our lives. He pushed us put positive things in our minds.

This concept of how thoughts become things is not a new concept. You may have read about it in books like Think and Grow Rich, or the Law of Attraction. Even many of the authors in the Bible remind us to think and focus only on good things and good things will happen.

One of the main thread of this blog Becoming Faithful is all about being intentional about life. Questions about the outcome of living with intentionality would fit here.

Why is it when we are intentional about something we see results? And, when we are not intentional we don't see results (or at least the results we were aiming for)

Some simple mind and matter questions most people will understand are:

- What is consciousness?
- What is intentionality?
- What is it for a mental state to be 'about' something?

They may not know the answer to these questions, but their eyes won't glaze over when you talk about them.

Furthermore, things like the subconscious mind, the universal mind would fit here. All vast topics.

9.6 RELIGION AND SPIRITUALITY

This broad topic of religion and spirituality is what lead me to learn all about metaphysics.

The area of topics within religion and spirituality and metaphysics are quite common. People should know what you're talking about when you talk about things like:

- Is there an afterlife
- Arguments for or against the existence of God.
- The problem of evil
- Divine grace
- All religions are one
- Souls
- Miracles
- There are many more areas under religion and spirituality. But, these are the key ones that people would have either asked themselves or have heard before.

I'm confident at one point or another, someone has asked if there is an afterlife. Or, questioned the existence of God.

9.7 METAPHYSICS OF SCIENCE

Metaphysics of Science is a subdiscipline of philosophy concerned with philosophical questions that arise at the intersection of science, metaphysics, and the philosophy of science. The term "Metaphysics of Science," which combines the names of these disciplines, is of 20th century coinage. In order to fully understand what Metaphysics of Science is, it is helpful to clarify how it differs from both metaphysics simpliciter and philosophy of science.

9.7.1 Metaphysics and Metaphysics of Science

Metaphysics simpliciter seeks to answer questions about the existence, nature, and interrelations of different kinds of entities—that is, of existents or things in the broadest sense of the term. It enquires into the fundamental structure of the world. For example, it asks what properties are, how they are connected to the entities which have them, and how the similarity of objects can be explained in terms of their properties. The subject matter of metaphysics is somewhat heterogeneous: topics include the composition of complex entities, the identity and persistence of objects, problematic kinds of entities (that is, entities about which it is unclear whether or in what sense they exist at all, like numbers and fictional objects such as unicorns), and many more. Metaphysics is usually understood as working at an abstract and general level: it is not concerned with concrete individual things or particular relations but rather with kinds of things and kinds of relations.

Metaphysics of Science is not completely disjoint from metaphysics simpliciter. Not only does it draw on the pool of methodological tools employed in metaphysics, but there is also substantial overlap regarding subject matter. Metaphysicians have their own reasons, independently of science, to investigate causation, modality, and dispositional properties, for example. Like space and time, these concepts pertain also to everyday phenomena. Although Metaphysics of Science, too, is usually attentive to our everyday intuitions and opinions about such phenomena, it engages in a specific investigation of the roles these concepts play in scientific contexts.

Metaphysicians of science often take scientific realism for granted—that is, they hold the philosophical stance that the sciences are apt to find out what the world is really like, that they track the truth, and that the entities they postulate exist. Antirealism about science, on the other hand, often coincides with a skeptical or agnostic attitude towards metaphysics. In the context of some broader metaphysical inquiries, scientific endeavors might well be seen as but one way to the truth. A mainly science-guided metaphysics might even be seen as mistaken.

Moreover, metaphysicians of science demand of themselves that they pay attention to discourses within the sciences. For example, some physicists like Richard Feynman (1967) speak of fundamental symmetry principles and conservation laws as being constraints on other, less fundamental laws of nature, rather than being laws about what is going on in the world. Metaphysicians working to develop a philosophical theory of nomicity, therefore, should allow for the possibility of there being laws of nature as well as laws of laws.

In short, Metaphysics of Science is that part of metaphysics that enquires into the existence, nature, and interrelations of general kinds of phenomena that figure most prominently in science. Also, Metaphysics of Science grants the sciences authority in their categorization of the world and in their empirical findings.

In terms of content, the transition between Metaphysics of Science and science might well be smooth with no clear border, so the distinction might be one that can only be made sociologically, regarding the departmental structure of universities or focusing on the practitioners and their methods of inquiry. Whereas many physicists engage in experimental work, metaphysicians are happy merely to consult the findings of their empirically working colleagues from the science departments.

9.7.2 Philosophy of Science and Metaphysics of Science

On the other hand, Metaphysics of Science may just as well be called a part of the philosophy of science. Philosophy of science consists of the philosophical reflection on the preconditions, practices, and results of science in general and of the particular sciences. Many philosophers of science are engaged in debates surrounding science as a (putative) source of knowledge: what makes scientific results especially reliable? That is, what distinguishes science from non- or pseudoscience, everyday knowledge, and philosophy? Which kinds of methods do and should scientists employ? What is scientific progress? Are scientific theories true? Are we ever justified in advocating a particular scientific theory, given that most scientific theories of the past have been replaced by others (like, for example, Newtonian mechanics was replaced with relativistic mechanics)? Can the sciences be unified into one big Theory of Everything? Together, these questions constitute the epistemology of science, that part of the philosophy of science which studies scientific knowledge.

Metaphysics of Science complements the epistemology of science. Whereas the latter asks questions of the sort, "How do we know of x?" Metaphysics of Science enquires, "What is the nature of x?" where "x" is a placeholder for some (kind of) entity, state of affairs, or fact discovered or postulated by science.

The task of Metaphysics of Science is not simply to list these entities or facts. Rather, it operates at a higher level of abstraction. For example, whereas the particular sciences inquire into specific causal relations—or, differently put, into some particular relation that holds between two particular measurable quantities, like the concentration of a drug and the soothing effect it has on headaches—Metaphysics of Science attempts to say what causation is in general. That is, it asks exactly which features a relation must have in order to count as a causal relation (like regular occurrence or modal force), and what the respective relata are. In short, Metaphysics of Science enquires into the key concepts of science not at the empirical but at a more abstract and general level.

Explication

Philosophers disagree about which key concepts constitute the subject matter of Metaphysics of Science. Some argue for a narrow interpretation of the term and claim that Metaphysics of Science is primarily concerned with concepts which are relevant to all branches of science, because without these central concepts, science would not be possible. For example, they suggest that kindhood, lawhood, and causation are concepts of this kind. Others, for example the Society for the Metaphysics of Science, are more permissive: they also include in the domain of Metaphysics of Science issues that arise in only some branches of science, such as problems regarding species (biology), intentionality and consciousness (psychology), and social kinds. Probably due to the emphasis that 20th century philosophy of science placed on physics, the larger part of debates within Metaphysics of Science revolves around topics that occur most prominently within the realm of physics, but which figure or bear connections to the other sciences as well:

- Laws of nature, causation, and dispositions
- Necessity, possibility, and probability
- (Natural) kinds and essences

- Reduction, emergence, and grounding
- Space and time.

Regardless of whether philosophers defend a narrow or a more permissive notion of Metaphysics of Science, they agree that the concepts in question are in need of explanation. At the very least, such an explanation must show how the concepts cohere. Some metaphysicians take one or more of the concepts they discuss (alongside their related phenomena) to be primitive, meaning that these concepts cannot be analyzed in terms of other concepts and their related phenomena cannot be subsumed under other phenomena. Typically, they then proceed to show that other concepts (alongside their related phenomena) can be explicated in terms of these primitive concepts.

As a discipline in its own right, Metaphysics of Science is still relatively young, especially when compared to other areas of philosophy (such as epistemology and ethics). Its topics, however, are not. For as long as science has existed, there has been metaphysical reflection on central scientific concepts. Metaphysics of Science of the 21st century differs from natural philosophy of the past in that the aspiration of natural philosophy was to speculatively describe the world as it is, whereas Metaphysics of Science is more concerned with what the world would be like if our best scientific theories were to turn out true.

9.8 METAPHYSICS OF SCIENCE IN THE 20TH (AND EARLY 21ST) CENTURY

9.8.1 The Logical Empiricist Critique of Metaphysics

Of the many historical roots of modern philosophy of science, Logical Empiricism stands out. The Logical Empiricists and their sympathizers were the progenitors of a new kind of philosophy. They influenced many of the most prominent philosophers of the late 20th century. In a sense, it is with them and their themes (laws of nature, causation, counterfactuals) that modern Metaphysics of Science begins, although they would have rejected much that currently goes by that name. Their ideas sparked many of the debates central to Metaphysics of Science.

In the 1930s, the Logical Empiricists proposed an empiricist, positivist program. They held that experience is our only source of nondefinitional knowledge (hence Logical Empiricism) and that the task of philosophy is logical analysis; that is, analysis of the logical features of and relations between sentences (hence Logical Empiricism). According to the Logical Empiricists, all the empirical propositions we believe can be reduced to so-called protocol sentences, which are direct renderings of our perceptual experience, or "the given." Only if we know how a sentence could in principle be verified—that is, which possible observations would result in our accepting it as true—can we say that the sentence is meaningful. This so-called verifiability criterion of meaning has one purpose in particular, namely, to exclude metaphysical speculation from the realm of meaningful discourse. For example, the metaphysical sentence "every thing has an immaterial substance" cannot be empirically verified; hence, according to the verifiability criterion of meaning, it is meaningless. A radical antimetaphysical stance was one of the key tenets of Logical Empiricism. Note that verificationism recasts the Empiricists' epistemic doctrine that all factual knowledge comes from sense perception as a semantic doctrine. Indeed, if we believe that what we know is expressed (or at least expressible) in meaningful sentences, then the transition from Empiricist epistemology to semantics is straightforward: all factual knowledge is expressed in meaningful sentences and only those sentences for which we are able to give a method of verification in observation are meaningful.

It soon became apparent, however, that Logical Empiricism, and especially the verifiability criterion of meaning, houses some serious flaws. Two major blows came from Willard Van Orman Quine's seminal paper, Two Dogmas of Empiricism (1951), which argued that two assumptions the principle of verification has to presuppose are untenable: the first is that there is a clear distinction between analytically true and synthetically true sentences. The second is that each meaningful sentence faces the tribunal of sense experience on its own for its verification or falsification.

Logical Empiricism faces further problems. Clearly, the Logical Empiricists held the sciences in high esteem. Usually, it is taken for granted that the sciences aim to discover natural laws and that they research properties such as electro-conductivity of different materials, reactiveness of chemical compounds, and fertility of organisms. Prima facie, it seems that many laws of nature can be expressed as general statements, that is, as statements of the form "any particular thing x which has property F also has property G" (in logical notation: $x(Fx \rightarrow Gx)$). For example, we say that all samples of metal expand when heated. But universal generalisations of this kind cannot ever be proven true by actual empirical observations (because they have far more instances, maybe infinitely many, than could ever be observed and confirmed), so the verifiability criterion rules out (at least some) laws of nature as meaningless. Even if this consequence could be avoided, what the laws of nature say is often taken to not be merely accidentally true, but to ensue with modal force. Empirically, we cannot account for modality: we can only observe what is actually the case, not what else is possibly or necessarily true.

Similarly, Logical Empiricism runs into problems regarding dispositional properties. Everyday properties such as solubility and scientific properties like conductivity cannot easily be reduced to the observable qualities of soluble or conductive objects. For example, a sugar cube is a somewhat solid object, much like a matchstick, but if we were to place the sugar cube in water, it would dissolve, whereas the matchstick would not. Its manifest properties such as solidity, color, and taste provide no clue as to what will happen to the sugar cube if placed in water. What is more, even if a particular sugar cube were never placed in water at all (or if it were placed in water but the water was already supersaturated with sugar so that the sugar cube would not dissolve in that particular situation), it would nevertheless retain its dispositional property of being soluble, although there is nothing about it that we observe which hints at its solubility. An analogous case can be made regarding dispositional properties discussed in the sciences, like conductivity or chemical bonding propensity, and similarly, regarding science's theoretically postulated, not directly observable, entities like quarks or superstrings. Because dispositional properties, theoretical entities, and universally generalized laws of nature appear to belong to the conceptual inventory of the sciences, Logical Empiricism, which fails to adequately account for them, quickly became an unattractive option.

9.8.2 The Return to Metaphysics

The failure of Logical Empiricism to cope with some of the key concepts of science eventually led to the development of Metaphysics of Science. Philosophers realized that if concepts such as law of nature and necessity could not be eliminated by reduction to observation terms, it must then be legitimate to examine them thoroughly, by whatever means seem fit. The most likely candidate to fulfill this task is metaphysics.

The development of Metaphysics of Science occurred simultaneously with the revival of metaphysics in the analytic tradition of philosophy, a tradition that was rooted in Logical Empiricism (as well as in the linguistic turn, manifested by the ideal and ordinary language philosophies of the late 19th and mid-20th centuries). Analytic philosophers were initially hostile towards metaphysical questions. They rejected questions which transcended empirical observation or fell outside of the scope of the sciences. However, philosophers like Willard Van Orman Quine (most famously in his essay "On What There Is" (1948)) and Peter Strawson (especially in his monograph Individuals (1959)) soon realized that there is a supposedly innocent way of practicing metaphysics by describing human conceptual schemes rather than by speculatively conjuring up grand metaphysical edifices. Instead of laying claims to knowledge of the unobservable, they focused on finding out how humans in fact conceptualize reality—in their everyday language (Strawson) or their scientific theories where, if stronger authority is given to the sciences, the latter may revise the commitments of the former. Quineans favor the revision and are, hence, closer to the attitude of Metaphysics of Science, where Strawsonians give much credibility also to folk's general metaphysical background assumptions.

Encouraged by the failure of Logical Empiricism and the fact that metaphysical questions were once again beginning to be the subject of philosophical discussion, philosophers developed a renewed interest in metaphysics. They gradually grew confident in talking not merely about observations, semantics, and language, but also about reality.

Another significant step towards the return to metaphysics was the development of modal logic. Begun by Carnap—for example, in his Meaning and Necessity (1947)—the logic of necessity, possibility, and counterfactuality was refined considerably by Ruth Barcan Marcus (1947), Saul Kripke (1963), and David Lewis (1973a). Later, with Kripke's Naming and Necessity (1980) and Hilary Putnam's "The Meaning of 'Meaning'" (1975), the formalisms were given ontological interpretations and the belief in necessity in nature gained new justifications. Building on these developments further still, even (Aristotelian) essences saw their revival: see Kit Fine's work (1994) and its application within Metaphysics of Science by, for example, Brian Ellis (2001) and Alexander Bird (2007).

The return to metaphysics in the 20th century was not merely a trailblazing event for the development of modern Metaphysics of Science; rather, the two evolved alongside each other. For example, when it became acceptable for metaphysicians to speak of necessities in nature and discuss statements like "Water is necessarily H_2O," this paved the way for a realistic reading of other modalities, like nomological necessity or counterfactuality. These are, as we will see (in section 4b and 4c), central notions in debates on the nature and status of laws of nature in Metaphysics of Science.

9.8.3 Naturalized Metaphysics and Inductive Metaphysics

In the early 21st century, some philosophers argued for a naturalization of metaphysics. Their argument typically rests on the fact that the sciences appear to surpass metaphysics in many respects. The sciences, they claim, have a shared stock of accepted theories, a pool of respected methods and institutionalized standards, and they have predictive and technological successes to show for themselves. In contrast, there is long lasting dissent over positions and methods in metaphysics that rarely ever gets dissolved, and it is unclear what would even count as criteria for metaphysical success. As some metaphysical questions—such as "What is the world ultimately made of?" and "What is life?"—also belong to the domain of the sciences (physics and biology, respectively), naturalists insist that we must draw upon scientific findings to properly answer them.

Naturalistic metaphysicians come in all shapes and sizes. Some naturalists wish to prohibit any metaphysics that is not scientifically evaluable. Some suggest that we should take our clues from scientific practice. For example, Tim Maudlin (2007) argues that lawhood is primitive, as working scientists see no need to analyze the concept. Others still allow for the possibility of relevant questions which may not have straightforwardly scientific answers. For example, consider the question "What is it for a thing to persist through time?" Imagine we take a ship out to sea and, little by little, replace every single part of it until none of the original parts remain. Certainly, science can describe how the ship changes, but it will not tell us whether the ship we sail home is still the same as the ship that put out to sea. The latter becomes a pressing, genuinely metaphysical problem, especially when we ask an analogous question about a person's change and persistence through time.

What is important to remember is that although a naturalized metaphysics may, in a sense, also be called a "Metaphysics of Science," its proponents may have a very different sort of metaphysics in mind than that presented in section 4.

In the 21st century, some philosophers have stressed that Metaphysics of Science could well be an inductive/abductive enterprise that, just as the sciences do, generalizes empirical data and builds explanatory models on that basis (Interestingly, precursors of the idea of an inductive/abductive metaphysics developed simultaneously with Logical Empiricism) If so, metaphysical hypotheses might turn out to be fallible, only approximately true, and contingent.

9.9 WHY DO WE NEED METAPHYSICS OF SCIENCE?

In section 1 it was said that Metaphysics of Science examines the key concepts of science. But why do philosophers even bother to argue over issues in Metaphysics of Science? Is it not relatively clear what the basic concepts in science are and what they mean? Surely scientists know very well what they mean to say when they talk about the solubility of sugar, the second law of thermodynamics, and the relativity of space-time?

What inspires Metaphysics of Science is, of course, the idea that there is more to know about these phenomena and the concepts involved than science can say. Think of causation, for example. The concept of causation is commonsensical: we encounter causal processes in everyday life, like when we hit a golf ball with a putter and the ball begins to move, or when we drop a glass and it shatters. We intuitively distinguish these causal processes from noncausal processes. For example, if somebody in the next room sneezes as you raise your arm, you just know that raising your arm was not the cause of the other person's sneezing. Still, it is quite complicated to say what establishes a causal connection between two events and what exactly distinguishes the putter-and-golf-ball scenario from the raise-arm-and-sneeze incident. Science records measurements and reveals statistical correlations between phenomena. It also has apt intuitions about whether two events are indeed causally connected or whether they merely co-occur accidentally, albeit regularly. Yet science is rarely interested in a general overall theory of what exactly distinguishes causes from accidents. Concepts such as causation or laws of nature, although relevant for science, are rarely the subject matter of science itself.

Science and Metaphysics of Science have different but complementary approaches to reality: the scientist's work in this respect is predominantly empirical and consists in finding instantiations—describing particular causal interactions, listing things which are disposed in certain ways, pinning down particular laws of nature, and so on—while the metaphysician's focus is on understanding and clarifying general concepts or the corresponding phenomena.

Still, the critic may object that even if the metaphysician's and the scientist's approaches to reality are indeed complementary, we can do perfectly well without Metaphysics of Science. For example, if science manages to find out the different variables and constants that determine how things in the world hang together, why do we also need to know what the general characteristics of a law of nature are or how that notion can be analyzed in terms of other notions? Isn't this superfluous information? Clearly, scientists do not need metaphysicians to tell them about causation or dispositions in order to perform their research. Nevertheless, metaphysicians of science believe that questions regarding the existence and nature of causation, natural kinds, and necessity are valuable in their own right. At the very least, they are pressing questions that cannot be ignored by those who yearn to thoroughly understand the world we live in. By way of example, consider the dispute between defenders of Humean supervenience and antiHumeans, which revolves around the question of whether there are necessities in nature or not. Clearly, this is not a question that can be answered by purely scientific methods, but it is one that metaphysicians will nevertheless take to be meaningful and profound.

Some of the issues discussed in Metaphysics of Science are also relevant for practical contexts. For example, failure to render assistance can lead to prosecution or social repercussions due to immoral behavior. However, you can only be held legally and morally responsible for events you are also causally responsible for. Accordingly, both ethics and law require a concept of causality that accounts not just for positive but also for negative causation, that is, causation by the absence of an event or act. If you pass an unconscious person lying on train tracks and fail to alert the authorities or pull him off the tracks, then you are (partly) causally responsible for his death if he is later killed by a train. Thus, although many questions within Metaphysics of Science are primarily aimed at complementing science, its debates may have far-reaching consequences in other fields as well.

To more fully understand the difference between the scientific and the metaphysical approach to the key scientific concepts that constitute the subject matter of Metaphysics of Science, it is helpful to consider samples of actual work in Metaphysics of Science (section 4) and to take a closer look at the methodology employed (section 5).

9.9.1 Sample Topics in Metaphysics of Science

As Metaphysics of Science is the study of the key concepts of science, its subject matter depends directly on what the sciences study and which concepts they employ. Because there are many different branches of science, there are also many potential topics for metaphysicians to discuss. It is impossible to name them all in a survey article, much less discuss them in detail. However, it is practically impossible to fully grasp what Metaphysics of Science is from general definitions only.

In order to give the reader an idea of both the scope of Metaphysics of Science and its practice, this section briefly and tentatively introduces seven debates which have preoccupied metaphysicians of science in the past: counterfactuals and necessities, dispositions, laws of nature, causation, natural kinds, reduction and related concepts, and space and time.

Dispositions

Some objects have dispositional properties. For example, sugar is soluble, matchsticks are inflammable, and porcelain vases are fragile. Properties like solubility or fragility are often conceived of as becoming manifest only under so-called "triggers" or "stimulus conditions," which set off the manifestation of the dispositional property. For example, for a sugar cube to manifest its solubility by dissolving, it must be placed in water.

Not all properties are like that. So-called categorical properties need no stimulus; they are always manifest. Just think of the properties of being solid, having a certain molecular structure (for example, being H_2O), being rectangular, and so on. The distinction between categorical and dispositional properties is often drawn with the following three features in mind:

(i) Untriggered dispositions are not directly observable, whereas many categorical properties are. For example, from looking at some sort of powder, we cannot tell whether it is soluble or not. Looking at a football, we immediately see that it is round.

(ii) Because dispositional properties bestow objects with possibilities (of behaving in certain ways under certain circumstances), they are said to be modal properties: they imply, by their very nature, what can, might, or (given certain circumstances) must be the case. Categorical properties are not usually conceived of in this way.

(iii) Dispositional properties are often identified with productive powers. For example, scratching a match is not enough for it to light up; the match's inflammability, too, is causally responsible for the flame. Usually, no such productive, causal force is directly associated with categorical properties.

Dispositional properties are not just a phenomenon we encounter in everyday contexts, but in science as well. For example, the property of being charged appears to fit this profile: it is not directly observable, it determines how objects would behave under certain conditions, and an object's charge can be a vital factor in causal processes. Dispositionality has hence been of interest to Metaphysics of Science since its very beginning. In fact, the failure of Logical Empiricism to properly account for dispositional properties played a seminal role in the emergence of the discipline.

Because of their shared belief that all of our knowledge ultimately reduces to observational experience, Logical Empiricists like Rudolf Carnap (1936) attempted to account for dispositional properties in terms of observational properties using a simple conditional to connect the trigger to the manifestation: to say that a sugar cube is soluble just means that if we put it in water, it will dissolve. This and similar attempts at reduction fail, however, as they do not account for the modal behavior of disposed objects. For example, they do not supply a basis on which to ascribe (or not to ascribe) solubility to objects which have never been placed in water. This strikes us as odd, as it does not correspond to our everyday practice.

In order to adequately capture the modal nature of dispositions, philosophers soon suggested that we employ a counterfactual connective instead of the simple conditional. To say that some object has a disposition, they argued, means just that if the object were exposed to the trigger conditions, the disposition would manifest. This approach faces at least two problems. First, it requires a theory that specifies truth conditions for counterfactual conditionals. Second, there are some interesting counterexamples to the effect that under certain conditions we would intuitively ascribe dispositions to objects for which the proposed analysis fails.

Although early attempts at reducing dispositions to categorical properties have failed, problems like the above have convinced some philosophers that we should strive for a reductive analysis after all. The philosophical position that holds that all properties are categorical and that supposedly dispositional properties can somehow be reduced to categorical properties is called "categoricalism." For many categoricalists, a large part of their motivation comes not from Logical Empiricism but a fundamental insight of classical empiricism. David Hume famously observed that necessary connections, like those between causes and their effects, cannot be detected empirically. Hence, Hume concludes, we have no reason to assume that any sort of productive, necessary, or modal connection of events in nature exists. Twenty-first century Humeans, too, claim that there are no necessary connections in nature. Consequently, they deny that there are irreducible, metaphysically fundamental dispositional properties that seem to imply some sort of necessary or modal connection between the trigger and the manifestation.

However, as reduction proves to be notoriously complicated, other philosophers opt for dispositionalism instead, which is, in its most radical form (pan-dispositionalism), the view that all properties are of a dispositional nature. Both categoricalism and pan-dispositionalism are monistic theories, as both claim that there is, at the fundamental level, only one type of property. It is also possible for philosophers to hold a neutral or dualistic view, according to which there are both categorical and dispositional properties at the fundamental level of reality.

The debate over dispositions has had substantial impact on other debates within Metaphysics of Science and vice versa. For example, some philosophers argue that laws of nature and causation are grounded in dispositional properties: a law of nature like "Like-charged objects repel each other" could well be true because of the dispositional nature of charge, and causal successions of events could be determined by the dispositional properties of objects involved. Other philosophers see the direction of dependence exactly the other way around: dispositions depend on laws of nature, because if the laws of nature were different, objects might have different dispositions. For example, if the laws of ionic bonding were different, salt might not dissolve in water. Similarly for causation: maybe salt has its disposition to dissolve because its ionic structure is a potential cause of dissolving. Hence, the debate over dispositions should not be viewed in isolation.

9.9.2 Counterfactuals and Necessities

We learned above that a central feature of dispositions is that they establish a modal relationship between the disposed object's being in the trigger condition(s) and the disposition's manifestation. A plausible candidate for understanding the nature of this modal relationship is counterfactual dependence. The standard notation for counterfactual dependence reads p → q: if p were the case, then it would be the case that q. If a sugar cube is soluble, then that means, at least in part, that if it were placed in water, it would dissolve.

The sentential connective → is an intensional connective, which means that the truth value of the entire conditional cannot simply be read off the truth values of the antecedent and the consequent. The reason is easily understood: counterfactual conditionals describe counterfactual situations, which means that both the antecedent and the consequent are usually not currently true. Yet some such counterfactuals with a (currently) false antecedent and a (currently) false consequent are true and some such counterfactuals are false. How then can we evaluate the truth of counterfactual conditionals, given that the truth or falsity of its components is not decisive?

An idea proposed by Nelson Goodman (1947, 1955) and Roderick Chisholm (1946) is to have the truth of a counterfactual conditional depend on both the laws of nature and the background conditions on which they operate. On this account, a counterfactual conditional p → q is true if and only if there are true laws of nature L and background conditions C which hold, such that p, L, and C communally imply q. (Some further conditions must be met, like that the background conditions must be logically compatible with p.) Obviously, if the laws of nature or the background conditions were different, p → q might turn out not to be the case.

An alternative way of thinking about counterfactuals called "possible world semantics" was introduced by David K. Lewis (1973a). Lewis's most important tool is the concept of a possible world. According to Lewis, our actual world is only one among a multitude of possible worlds. A possible world is best thought of as one way (of many) the actual world could have been: all other things being equal, the word "multitude" in the last sentence could have been misspelled, Lewis could never have been born, or atoms could have been made of chocolate. Robert Stalnaker (1968) proposed a similar account but without defending modal realism. To him, possible worlds are tools, and as such no more than descriptions of worlds that do not exist.

Some possible worlds are more similar to ours than others. For example, a world which is like ours in every respect except that "multitude" is misspelled in the preceding paragraph is more similar to the actual world than a world with chocolate atoms. In evaluating a counterfactual's truth value, this fact plays a seminal role. Consider, for example, the sentence "If David had not overslept, he would not have been late for work." In a world where all vehicles miraculously disappeared that morning, where the floor of David's bedroom was covered in super strong instant glue, or where the laws of nature suddenly changed so that movement is no longer possible, he would not have made it into work in time, even if he had gotten up early. But these worlds do not interest us; this is clearly not what we mean by saying that had David not overslept, he would have made it in time. To judge whether the counterfactual conditional is true regarding our world, we need to consider only worlds where the laws of nature remain the same and everything else is rather normal—that is, similar to what actually did happen—except for the fact that David did not oversleep.

Lewis and Stalnaker suggest that an ordering of worlds with respect to similarity to our world is possible. Naturally, worlds where many facts are different from the facts of our world, and worlds with different laws of nature, count as particularly dissimilar. Counterfactual truth can then be determined as follows: of all the possible worlds where p is the case (for short, the p-worlds), some will be q-worlds and others non-q-worlds (that is, worlds where q is true or not true, respectively). To determine whether the counterfactual conditional p → q is true for our world, we need to check whether the p-worlds that are also q-worlds are more similar to our world than the p-worlds that are non-q-worlds. So to find out whether it is true that David would have gotten to work in time had he not overslept, we look at possible worlds where David did not oversleep and check whether the worlds where he makes it into work are more similar to the actual world than worlds where he does not.

According to this analysis, the consequent need not be true in all possible worlds (but only in similar p-worlds) in order for a counterfactual to be true. For example, had David overslept in a world where objects can be transported via beaming, he might still have made it to work in time. But as it is doubtful whether this technology will ever be available in our world (as it is not clear whether it is compatible with our laws of nature), the world where beaming has been invented is not relevant for the evaluation of the counterfactual conditional.

Related to what has just been said, we can point out a welcome feature of counterfactual conditionals: it can be true both that if David had not overslept, he would not have been late for work; and that if David had not overslept, yet the bus had had an accident, he would (still) have been late for work. This is a feature that necessary conditionals and mere material implications cannot well accommodate.

In addition to providing a way of understanding counterfactual conditionals, possible world semantics allows us to spell out the modal notions of necessity and possibility in terms of quantification over possible worlds. Thus, a sentence p is necessarily true if and only if it is true in all possible worlds. If p is necessarily true, there is no way that p could be false; that is, there is no possible world where p is false. Similarly, p is possibly true if and only if it is true in at least one possible world.

Necessity is thus expressed in terms of universal quantification over (all) possible worlds, whereas possibility is existential quantification over (all) possible worlds. Like the general and existential quantifiers, necessity and possibility, too, are interdefinable: if p is necessary, then it is impossible that non-p, and if p is possible, then it is not necessarily the case that non-p.

Note that there are different sorts of necessity which can be easily accounted for if we conceive of necessity and possibility in terms of quantification over possible worlds: Logical, metaphysical, and nomological necessity can be defined by restricting the scope of worlds over which we quantify. For nomological necessity, for example, we restrict quantification to all and only worlds where our laws of nature hold.

Possible world semantics faces several problems, however. For example, it is unclear just how we can know about what is or is not the case in other possible worlds. How do we gain access to possible worlds that are not our own? However, possible world semantics is a valuable tool for understanding some of the most central issues in Metaphysics of Science, such as dispositions and causation. In addition, necessity is a crucial element in theories of laws of nature, essences, and properties. The modalities of necessity, possibility, and counterfactuality are also important in their own right: after all, knowing what would happen if something else were the case or what can or must happen is key to scientific understanding.

9.9.3 Laws of Nature

Here are some intuitions philosophers have about laws of nature: laws are true or idealized, objective, universal statements. Laws of nature support counterfactuals, are confirmable by induction, and are explanatorily valuable as well as essential for predictions and retrodictions. Laws have modal power in that they force certain events to happen or forbid them from occurring. Any analysis of the concept will attempt to account for at least some of these features. Roughly, there are five types of theories of laws of nature: regularity accounts, necessitation accounts, counterfactual accounts, dispositional essentialist accounts, and accounts which take laws to be ontological primitives.

The basic idea of early regularity accounts is that a law of nature is a true, lawlike universal generalization (usually of the form "All F are G," or in formal notation: x (Fx→). Whether a given generalization is true is, of course, an empirical matter and must be determined by the sciences, but what it means for a statement to be lawlike is left for metaphysics to define. Not all general statements are lawlike. For example, some general statements state logical truths which clearly are not laws of nature. The main challenge for regularity theories is figuring out what makes a universal statement lawlike without appealing to any sort of connection between events other than regularity.

The Best Systems Account is an example of a sophisticated regularity theory. It asks us to imagine that all facts about the world are known, such that you know of every space-time point what natural properties are instantiated at it. There are many different ways of systematizing this knowledge by using different sets of generalizations. These generalizations make up competing deductive systems. Defenders of the Best Systems Account hold that a (contingent) generalization is a law of nature if and only if it is a theorem within the best such system. Which system is the best is determined by appeal to certain criteria: simplicity, strength (or informational content), and fit.

The Best Systems Account has been criticized for not taking seriously the intuitions that laws of nature are objective, have explanatory value, and hold with modal force. The Best Systems Account yields regularities, but it does not explain why they obtain. Opponents of regularity theories stress that laws do not merely state what is the case, but enforce or produce what happens.

Necessitation accounts are alternatives to the Best Systems Account that endorse this idea. Such accounts have been proposed by David Armstrong (1983), Fred Dretske (1977), and Michael Tooley (1977). For Armstrong, a law of nature is a necessitation relation N between natural properties. (Armstrong speaks of universals.) For two natural properties to be related by necessitation means that one of them gives rise to and must be accompanied by the other (hence necessitation). To give a coarse-grained example: Coulomb's law (which states, very roughly, that charges exert forces onto other charges), is a true law statement if and only if necessitation holds between the properties of having a certain charge (C) and exerting a certain force (F): N(C, F).

Necessitation accounts have some advantages over regularity theories. For example, they can more easily allow for uninstantiated laws. But how exactly do we know which properties are related by the necessitation relation, and why should we even assume that it exists? Armstrong argues that necessitation can be experienced insofar as it manifests in causal processes. However, not all laws are causal laws. Defenders of necessitation accounts must work out these issues.

The counterfactuals account focuses on a feature related to necessity, namely, the fact that laws of nature are stable under counterfactual perturbations. For example, that nothing can be accelerated beyond the speed of light is a law of nature because it is a fact that no matter what fantastical interventions we were to devise, we still couldn't travel faster than the speed of light. Versions of the counterfactuals account of laws of nature have been proposed by James Woodward (1992), John Roberts (2008), and Marc Lange (2009).

A bullet that counterfactual accounts have to bite is that the intuitive order of explanation regarding laws and counterfactuals is upside down: whereas the counterfactual theory of laws says that it is a law that all bodies fall down to earth because it is fundamentally true that "were some arbitrary massive body dropped it would fall," we intuitively believe that "were we to drop this body it would fall" is true because the law of gravitation holds. In other words, it is more intuitive to hold that the laws of nature support counterfactuals rather than that counterfactuals support the laws.

Another prominent way to account for laws of nature is to appeal to dispositional essentialism. Dispositionalists, like Brian Ellis (2001), Alexander Bird (2007), or Mumford and Anjum (2011), believe that some or even all properties are essentially dispositional. For example, if an object has the property of being electrically charged, that just means that it has the dispositional property of being attracted or repelled by other charged objects nearby. In this sense, the property of being electrically charged is essentially dispositional, because no object is electrically charged unless it is disposed to be attracted or repelled in this way.

Now, if natural properties bestow on their bearers dispositions, then that means it is always true that if something has a given natural property (Px), it also has a certain disposition (Dx) and thus it will manifest in a certain way (Mx), given that the disposition's corresponding trigger occurs (Sx). This is precisely what many metaphysicians ask of laws: that they bring about or make necessary what happens when something else is the case. Dispositional essentialists thus claim that dispositions ground nomological facts: laws arise from the dispositions things have.

Obviously, the dispositional essentialist account of lawhood hinges on non-trivial premises, which must be evaluated in their own right—for example, the premise that dispositions are basic.

If analyzing lawhood is so complicated an affair that it requires elaborate theories and intricate tools, why not assume that lawhood is conceptually and ontologically primitive—that is, that the concept of lawhood cannot be defined in terms of other concepts, and that it cannot be reduced to underlying phenomena? Tim Maudlin (2007) argues that scientists do not seek to analyze laws, but rather accept their existence for a brute fact in their daily practice, and that philosophers should do likewise.

To Maudlin, a law of nature is that which governs a system's evolution through time and determines what future states can be produced from the current state of the system. As lawhood is a primitive concept for Maudlin, he attempts to utilize it in defining other notions, like causation and counterfactual truth. Whether Maudlin's approach is viable or not depends to a large part on whether these definitions of causation and counterfactual dependence by means of laws of nature work out or not.

Causation

Causation is obviously intimately connected to the laws of nature, as we would expect at least some laws to govern some causal relationships. Causation, however, is not a straightforward notion. For example, philosophers disagree over which kinds of entities are the proper relata in causal relationships, some potential candidates being substances, properties, facts, or events. There are several approaches to understanding causation: regularity theories, counterfactual theories, transfer theories, and interventionist theories.

Regularity theories follow in the footsteps of David Hume's treatment of causation. According to regularity theories, all that can be said about causation comes down to stating a regularity in the sequence of events. The motivation for regularity theories stems from the fact that instances of a regularity can be observed, unlike the production of one event by another or a necessary relation between events.

One of the most widely known regularity theories is John Mackie's INUS account of causation (1965). According to Mackie, an event is a cause if it is an Insufficient but Necessary part of an Unnecessary yet Sufficient condition for the effect to occur. For example, a short circuit (C) alone is not sufficient for a house to burn down (E); there must also be inflammable materials nearby (A) and there must not be sprinklers which extinguish the fire (B). Call this a complex condition (ABC). As the absence of sprinklers and the presence of inflammable materials is not enough to cause a fire, the short circuit is necessary within this complex condition, which is then sufficient for the fire. But there may be other complex events (DFG, HIJ, and so on) which could also bring about the same effect. For example, a lit candle in a dried-up Christmas tree may also cause the house to burn down. As the short-circuit scenario (ABC) is only one of many potential causes of a fire, it is not necessary for the effect to occur, but if it occurs, it is sufficient to bring about the fire.

Like other regularity theories, Mackie's INUS theory has the disadvantage of classifying as causal some regularly co-occurring coincidences that are, for all we know, not causally related. For illustration, consider a simpler type of regularity theory according to which causation is just regular succession. The problem is that if causation were nothing but regular succession, then we would be forced to say that the rise of consumer goods prices in the late 20th century causes the oceans' water levels to rise. Obviously, these events coincided but are not causally related.

To forgo this problem, philosophers devised counterfactual theories of causation. The initial idea presented by David K. Lewis (1973b) is to equate causal dependence with counterfactual dependence. The idea seems plausible: had the cause not occurred, there would (all else being equal) not have been the effect. More precisely, for event e to (causally) depend on event c, whether e occurs or not must depend (counterfactually) on whether c occurs or not. For example, if the short circuit is the cause of the fire, then the house would have burned down if the short circuit had occurred, and it would not have burned down if the short circuit had not occurred.

Lewis saw that this initial account is flawed as it yields intuitively incorrect results in so-called pre-emption scenarios. Imagine two people, Suzy and Billy, throwing stones at a bottle. Now picture a situation where if Suzy does not throw her rock, Billy will. Suppose Suzy throws her rock, hits, and the bottle shatters. The effect, namely the shattering of the bottle, is evidently caused by Suzy's throwing the rock. However, the effect would have occurred even if Suzy hat not thrown, because in that case Billy would have thrown his rock and shattered the bottle. In this scenario, we recognize Suzy's throw as the cause of the shattering, but the latter does not counterfactually depend on the former.

Although more sophisticated counterfactual theories are more successful in dealing with pre-emption and other problems, some philosophers choose to take a different approach.

Proponents of transfer or conserved quantity theories like Salmon (1984, 1994), Phil Dowe (1992), and Max Kistler (2006) claim that causation is best understood as a transfer of a physical quantity from one event to another. For example, Suzy is causally responsible for shattering the bottle (and Billy is not) if it was her energy that set the stone in motion to physically interact with the bottle on impact and shatter it. Transferable quantities include energy, momentum, and charge, for example. These quantities are subject to conservation laws, which means that in any isolated system, the sum total of the remaining and the transferred amount of the quantity will always equal the initial amount.

Transfer theories face difficulties in accounting for negative causation. For instance, omitting to water plants may cause them to wither, but there is no transfer of a conserved quantity from anything to the withering. Other problems derive from examples where the supposed causal relationship is not obviously of a physical nature. For example, we may say that wild speculations at the stock market caused the economy to break down or that Suzy's throwing Billy a kiss causes him to blush.

Of the fourth group of theories of causation, interventionist theories, James Woodward's approach (2003) is a prime example. Woodward suggests that causation is best characterized by appeal to intervention. Consider the following example: Testing a drug for efficiency consists in finding out whether a group of people who are administered the drug are cured while a group who does not receive the drug remains uncured. In other words, drug testers intervene by giving the drug to some patients and a placebo to others. If the drug intervention leads to recovery while the placebo intervention does not, the drug is said to be causally relevant for the recovery.

Woodward places further constraints on interventions, one of which is that the intervention (of administering the drug or the placebo, respectively) must be performed in such a way that other potential influences are absent. For example, if the drug were given to healthy and young patients while only the elderly and frail receive the placebo, the test might falsely attribute causal efficacy to the drug.

Even when these precautions are taken, Woodward's theory is at risk of being circular: the analysis presupposes that we understand beforehand what it means to intervene on a system. Intervention, however, is itself a causal notion. Woodward has clarified that his theory is meant to explicate and enlighten our concept of causation, not to reduce causation to other phenomena.

It seems that all theories of causation face difficulties (either in the form of recalcitrant exemplary cases or in that they do not capture certain features of causation). One possible conclusion to draw from this is that causation is not one unified phenomenon but at least two and potentially many more. For example, Ned Hall (2004) argues that our intuitions characterize causation both as production and counterfactual dependence, and that the problems of analyses of causation can be traced back to the attempt of squeezing these into one unified concept.

The debates over the nature of dispositions, modality, laws of nature, and causation are still ongoing. Many promising approaches have been proposed in their course and will continue to be explored in the future.

9.9.4 Natural Kinds

In everyday contexts we habitually classify objects or group them together. Some of these groupings seem more natural to us than others. Philosophers who believe that nature comes with her very own classifications speak of "natural kinds." For example, samples of gold closely resemble each other, differ clearly from other chemical elements, and share a common microstructure, whereas sea life comprises organisms of very different sorts (including crustaceans, fish, and mammals). Terms like "sea life" and "tile-cleaning fluid" are convenient for human purposes such as thinking and talking about groups of things, but we do not expect them to reflect the structure of the natural world (which does not mean that the classifications they introduce are entirely arbitrary). Natural kinds, on the other hand, supposedly "carve nature at the joints". They are also highly projectible: we can inductively infer from the behavior of one object to that of all objects of the same natural kind.

If natural kinds exist and contribute to the structuring of the world, then ideally we want the sciences to discover what natural kinds there are. A natural kind enthusiast may claim that physics tells us that electrons and quarks exist, chemistry says that there are chemical elements like gold (Au) and compounds like water (H_2O), and biology seems to suggest that organisms are ordered hierarchically along the lines of family, genus, and species. However, there are also conventionalists who believe that so-called natural kinds are not independent of the minds, theories, and ambitions of human beings, or that no way of dividing up the world is inherently better than any other. To illustrate their claims, they remind us that the concept of biological species used to be regarded a prime example for natural kinds, but that, in the meantime, various paradigms have been proposed, each leading to a different system of classifications.

If natural kinds exist in nature, then what are they? What makes a natural kind the kind it is? Different ideas have been proposed and have given rise to a multitude of questions: Do objects which belong to natural kinds share at least some properties? Are these special, "natural" properties? Are natural kinds determined by the roles they play in inductive inferences or laws of nature? Is there a hierarchy of natural kinds, such that some kinds are more fundamental than others?

A position that has been particularly influential in the 20th century is the view that natural kinds have essences. It supposedly follows from Hilary Putnam's Twin Earth thought experiment (1975). Suppose there is a planet just like Earth in every way, but there is a liquid that the inhabitants of Twin Earth call "water" and which resembles water in every respect except for its microstructure, which is not H_2O, but XYZ. Intuitively,

Putnam claims, XYZ is not water, which leads him to assume that, unlike the superficial properties of being wet, potable, and so on, being H_2O is a necessary condition for being water. Similar conclusions can be drawn from Saul Kripke's argument that if we were to find out that the color we have up to now associated with elementary gold is actually an illusion, we would all agree that gold remains gold so long as it has atomic number 79, no matter what color it is (1980).

Linked to but distinct from the question of what natural kinds are is the question of whether natural kinds form an ontological category in their own right, or if they can be reduced to other existents like properties. Realists regarding natural kinds believe that talk of natural kinds and successful inferences presupposes the existence of natural kinds in nature. Reductionists, on the other hand, may argue that membership in natural kinds is not only determined by a number of shared properties, but also that it consists in nothing over and above having these properties.

Unsurprisingly, metaphysicians of science are especially interested in finding out which, if any, natural kinds are postulated or discovered by the various branches of science and whether they really identify as natural kinds by the standards of contemporary metaphysical theories, or whether the theories of natural kinds need to be revised.

9.9.5 Reduction, Emergence, Supervenience, and Grounding

The world consists of many different things. Philosophers have always dreamed of rendering it more orderly by systematizing it in just the right way. An important step towards doing so seems to entail an analysis of the relationships and dependencies between things which belong to different strata or levels of reality. The world apparently comes structured in levels, with things on higher levels somehow depending on the things on lower levels. For example, a factory consists of machines, conveyor belts, and so forth; machines are made of various interacting cogs, levers, and wires; the cogs are made out of molecules, the molecules are made of atoms, and the atoms are made of protons, neutrons, electrons, and so on. Dependencies like these are studied by the various special sciences. It is clear, however, that a factory is not composed of machines, conveyor belts, and so on in the same way that an atom consists of particles. Surveying the whole of science, Metaphysics of Science strives to account for the various ways higher level objects depend on lower level entities. The aim is not just to establish what depends on what, but to also clarify and explicate the nature of the dependencies. The kinds of relations most fervently discussed in Metaphysics of Science include reduction, emergence, supervenience, and grounding.

Reduction is often conceived of as a two-place, asymmetrical relationship to the effect that one thing is somehow made of, accounted for, or explained in terms of another thing. Typically, the reduced thing is conceived of as somehow less fundamental or less real, or even considered to be eliminated. Two types of reduction are relevant to Metaphysics of Science. First, there is reduction of one theory to another. For example, is it possible to express some theories of chemistry in terms of physical theories? If so, can all chemical theories be thus reduced? What about biological, psychological, and sociological theories? Second, reduction is sought between different sorts of entities or ontological categories such as phenomena, events, processes, and so on. Potential candidates include reduction of macro-level objects to molecules, atoms, and subatomic particles, reduction of properties to sets of objects which resemble one another, reduction of states of affairs to entities and properties (including relations), and reduction of the mental to the physical. The latter especially has been widely discussed in metaphysics. For Metaphysics of Science, claims of reduction pertaining to entities postulated by the sciences are of great interest, as are claims regarding reductive relationships between theories and their key concepts.

In a way, then, an armchair is reducible to its constituent parts: the fabric, upholstery, wood, and metal springs. However, an armchair is obviously not the same as a random pile of these materials. Unsurprisingly, philosophers disagree over whether, for particular cases, complete reduction can be achieved or not. Sure, a particular performance depends on the physical movements of the musicians and on how the created soundwaves causally impact on the hearers' eardrums, but the Concerto is not identical to these physical properties apparent in any given performance of it, as it exists independently of them.

Those who argue that such reductions do not succeed often speak of the irreducible as emergent from the underlying basis. They want point out that although there is a dependence of the higher on the lower level, the higher level adds something novel and can thus not be completely reduced to the lower level. An emergent property or phenomenon cannot be accounted for by reduction, because it is believed not to be a property of any of the component parts, and it is not obviously caused solely by their interplay. For example, whether or not you find abortion morally reprehensible does not seem to depend on the physical facts. Given the same situation, somebody else might pass the opposite moral judgment. Whereas such moral considerations are of no great professional import to the metaphysician of science, emergent properties in the sciences are. For example, biology still struggles to explain why higher forms of life have certain properties like consciousness, aspirations, and phenomenal experiences, which are not obviously properties of the underlying matter.

Reduction and emergence are interlevel relations. The most innocent, weakest dependence relation that is compatible with both reduction and emergence is called supervenience. Some thing A (the so-called supervenience set) is said to supervene on some other thing B (the so-called supervenience base) if and only if there can be no difference in A without there also being a difference in B—or, for short, if there is no A-difference without a B-difference. For example, an oil painting's macro-properties (A)—what it depicts and how it looks to us—supervene on its microphysical properties (B): unless the location, intensity, or color of the paint blotches are changed, the painting will always look the same to us. To better understand the world, metaphysicians of science research supposed supervenience relations in the sciences.

In the early 21st century, metaphysicians turned their attention to another sort of interlevel relation: grounding relations. Grounding relations are metaphysical relations which establish a special sort of (noncausal) priority of one over the other. Of two propositions or facts which are related by a grounding relation, one is taken to ground, or account for, the other. Grounding is stronger than supervenience, as it amounts not just to the claim that some A-facts only vary when B-facts vary—which may occur coincidentally—but that A-facts vary because B-facts vary. Unlike some forms of reduction, grounding does not seek to eliminate the grounded fact; attributing full existence to both of them, it merely ascribes a more fundamental status to the grounding fact.

Debates over grounding revolve around a number of pivotal questions, such as whether instances of grounding are all of the same kind or whether they embody a number of different relations, whether the grounding relation is primitive or can be analyzed in terms of other relations, and whether it is an irreflexive, asymmetric, and transitive relation or if other properties should be ascribed to it. The answers to these questions may also have an effect on how we should conceive of interlevel relations in the sciences, and the latter are of great interest to metaphysicians of science.

Space and Time

To most philosophers interested in the field, Metaphysics of Science is not confined to discussing concepts that pervade the whole of science (as, arguably, law of nature and causation do). It is also concerned with metaphysical questions that arise with respect to the particular sciences, like "What is life?" (biology) or "What is the ontological status of cultures, governments, and money?" (sociology). The philosophy of physics, too, gives rise to many interesting metaphysical questions. Among them are questions regarding the nature of space and time, which have been debated since the early dawn of western philosophy and, in the light of modern-day physics, are still at issue in philosophical debates.

As humans, we perceive space and time as different phenomena with differing properties. Space, as we perceive it, extends in three dimensions, and we can (almost) freely move in any direction. Through the physical forces which act upon our bodies, we are capable of detecting some sorts of motion through space (like when we run or jump) but not others (like Earth's rotation). Time, on the other hand, has a sort of directedness to it (commonly referred to as "the flow of time"). We cannot linger at a particular moment in time, and we cannot go back to previous times. Entities somehow change yet persist through time.

Metaphysicians of science are interested in these phenomena especially in the light of Albert Einstein's theories of Special and General Relativity. These theories were proposed in order to make sense of the fact that the speed of light was measured to remain constant regardless of the motion of the light source, whereas the velocities of objects depend on the motion of the object relative to an observer. For example, the speed of a train measured by a stationary observer on the platform is greater than its relative speed with respect to another, slower train that moves in the same direction. The speed of light emitted by a lamp on the train, however, will be the same regardless of whether it is measured by a passenger or a bystander. In popular interpretations, Einstein's theory of Special Relativity suggests that the problem can be solved by postulating that the three spatial and the one temporal dimension form a continuum by the name of space-time. An astonishing consequence could be this: Different observers are at motion with respect to different objects. Their perception of the present is determined by which information is accessible to them, which in turn is a matter of which light signals reach them at a given moment. Therefore, their individual present, past, and future differ according to their state of motion with respect to other objects. Thus, an objective, observer-independent order of points in time does not exist. This view is often referred to as the block universe view, because everything seems to simply exist conjointly, with no objective past or future. Some philosophers also suggest that, on this view, familiar material things are three-dimensional slices of four-dimensional objects.

Some philosophers claim that the block universe view is incompatible with presentism (the philosophical position that holds that only what is present exists) and supports eternalism (the view that all events past, present, and future exist). Unfortunately, the latter seems not to correspond to our subjective experiences of time. This poses a genuine dilemma for metaphysicians: should we accept Einstein's theories and dismiss our subjective experiences, or do we need to reinterpret the remarkably well corroborated theories to accommodate our everyday conceptions of space and time?

More such fascinating questions remain. How is the (perceived) directedness of time and its irreversibility (which manifests as increase of entropy) best explained? Are space and time finite or infinite? Do they exist fundamentally and independently of the objects in them, or does their existence hinge on the existence of those objects? Quite obviously, these are questions on which scientific theories have a bearing, and Metaphysics of Science works towards solutions that are both philosophically rewarding and scientifically tenable.

9.10 THE METHODOLOGY OF METAPHYSICS OF SCIENCE

Although Metaphysics of Science is concerned with the key concepts that figure prominently in science, its methods are not predominately those of the sciences. Apart from referencing scientific results and practices, Metaphysics of Science has a number of argumentative tools at its disposal that do not usually play an explicit role in scientific methodology but are not entirely unscientific either. In science these forms of arguments are implicitly employed to establish hypotheses when the empirical evidence is insufficient (for example, because two theories are equally well supported by the available evidence). Unlike many scientific theories, metaphysical claims often cannot be tested experimentally at all—not because we lack the technological means to do so, but because the very nature of these claims defies empirical confirmation or falsification.

Think, for example, of the claim that laws of nature hold across all possible worlds. This is why reference to theoretical virtues, Inferences to the Best Explanation, arguments from indispensability and serviceability, extensional adequacy, and the Canberra Plan method are of great argumentative importance in Metaphysics of Science.

Note that some philosophers—for example, proponents of naturalized metaphysics—may reject all or some of these methodological tools as transcendental or indefensibly a priori. However, the issue is not currently settled among philosophers, and the tools described below remain widely used in contemporary Metaphysics of Science.

9.10.1 Theoretical Virtues

In both science and metaphysics, we strive for internally consistent, comprehensive, unambiguous theories which cohere with our accepted beliefs, have an adequately large scope, and so on. Among the various desiderata, explanatory power and simplicity are often accorded a central role. To strive for an explanatorily powerful theory is to demand that a theory must explain a certain number of phenomena which stand in need of explanation, that it does so thoroughly and systematically, and that it is not ad hoc. The value of explanatory power is obvious: explanation (or at the very least, systematization) is the very purpose of any hypothesis. Not so with simplicity. There are many ways a theory can be-simpler than its competitors; for example, it may contain fewer variables than another. Usually, the call for simplicity is understood in terms of parsimony. Occam's Razor, a principle frequently appealed to in this context, says that entities must not be multiplied beyond necessity—that is, if faced with otherwise equally good theories (in terms of their explanatory power, for example), we are to prefer the one that postulates fewer entities. However, it is unclear whether simplicity and the other explanatory virtues are truth conducive or whether they are primarily pragmatic or aesthetic theoretical virtues.

Although theory choice criteria are certainly at work in everyday reasoning, philosophy, and science—remember that nobody wants a complicated, inconsistent, unclear, shallow, or incomprehensive theory—the application of such criteria is not straightforward: they must be measured and traded off against each other. Unfortunately, there are no shared standards or guidelines on how this should be done. How do we find out which of two theories is simpler or more consistent with the body of already accepted beliefs? How do we know which criterion trumps another? What is more, whereas in science theory choice criteria are interim solutions until a theory can be empirically proven, there is usually no such post hoc test in Metaphysics of Science. For all these reasons, justifying our appeal to theoretical virtues is not a trivial or easy task.

9.10.2 Inference to the Best Explanation

Once it has been determined through careful assessment of the theoretical virtues which available theory is the best explanation for a given phenomenon, we tend to infer that it must also be the correct explanation. In most cases, we will then also say that the entities (objects, fields, structures) postulated in the explanatory theory really exist. That is, we apply a so-called Inference to the Best Explanation (often referred to as "IBE"). For example, astronomers found that the best explanation for a divergence in the orbit of Uranus is the existence of another planet, Neptune, whose gravity interferes with Uranus' trajectory. Thus, they inferred that Neptune must exist. This hypothesis was confirmed when Neptune was later discovered through telescopes. Similarly, many metaphysicians of science believe that IBE can be applied to metaphysical theories. For example, Nancy Cartwright believes that the best explanation for the fact that laboratory results produced in controlled, sterile settings can be applied to the messy circumstances of the outside world is the existence of underlying dispositions that are examined in the laboratory but also pervade the rest of the world, and she therefore accepts this view as true.

Quite obviously, IBEs are not deductively valid, and even the best explanations we have at our disposal can later turn out to be incorrect. For example, when astronomers sought to explain anomalies in the orbit of Mercury, they failed to find Vulcan, a planet postulated explicitly for this purpose, and the anomalies were later explained with the help of the General Theory of Relativity.

Note also that Occam's Razor and IBEs sometimes pull in opposite directions: whereas IBEs often enrich, rather than reduce, our ontology, Occam's Razor is set on eliminating as many entities as possible from our ontology. On the other hand, one of the marks of a good explanation is that it does not postulate more than is necessary; that is, it is parsimonious in the sense of Occam's Razor. Either way, even if metaphysicians can agree on using theoretical virtues and IBEs as argumentative tools, there is still room for debate.

9.10.3 Indispensability and Serviceability Arguments

In addition to IBEs, metaphysicians appeal to further inferential arguments to the effect that we should accept certain hypotheses as true. More specifically, indispensability and serviceability arguments basically consist in claiming that if X plays a crucial role with respect to Y, and if Y is either uncontroversial or relates to some postulate that we are unwilling to let go, then the existence of X can (or must) be asserted—that is, we should believe that X exists for the sake of Y.

One reason for accepting the existence of an entity X may be that its existence is indispensable for the existence of Y; that is, Y cannot be the case unless X exists. For example, some metaphysicians argue that the existence of mathematical entities is indispensable for science, and as science is important and probably at least approximately true, we have every reason to believe in the existence of numbers. Very roughly put, indispensability arguments infer from the premise that X is indispensable for Y and the premise that Y is the case to the conclusion that X exists.

(An older variant of the argument from indispensability is the so-called transcendental argument, which usually runs like this: if X is a necessary condition for the possibility of Y, and if we believe that Y is the case, we should also hold that X exists.)

Serviceability arguments are weaker than indispensability arguments. They advise us to accept the existence of a (kind of) entity X if X is serviceable towards end Y. For example, David K. Lewis argues that the assumption that possible worlds are concrete objects (just as our actual world) is highly serviceable (1986, 3): among other things, it provides us with the means to spell out the semantics of counterfactual conditionals. However, there may be other ways of accounting for the truth conditions of counterfactuals (for example, by referring to complete descriptions of fictitious possible worlds instead). Whereas indispensability offers a strong argument for the existence of some sort of entity, serviceability allows for contenders. Different kinds of entities may serve equally well to implement a goal, and serviceability arguments alone may not suffice to determine which of these entities we should believe in.

The evaluation of indispensability and serviceability arguments depends on what you already believe and what goals you pursue (as represented by variable Y). At best, they yield conditional existence claims: if you believe that science is successful and that science would not be successful if it were not for the existence of mathematical entities, then you had better believe in the existence of mathematical entities. If you do not believe that science is successful, then the argument is moot. Awareness of the occurrences of these kinds of arguments within debates in Metaphysics of Science will certainly help you understand your opponent, but it will seldom suffice to settle the issue.

9.10.4 Extensional Adequacy and the Canberra Plan

One particularly useful tool in evaluating metaphysical hypotheses is the test for extensional adequacy. To test a theory for extensional adequacy means to examine cases that, according to pretheoretical, intuitive judgment, fall under a concept the theory aims to explicate and to check whether the theory indeed subsumes these cases as instances of the concept. In addition, the theory may be tested with regard to scenarios in which its concepts should intuitively not apply; if the theory (wrongly) applies, it may have to be corrected. For example, suppose someone proposes a metaphysical theory as to what a law of nature is in claiming that a law of nature is nothing but a general statement of the form "All things which have property F also have property G." This theory will quickly be challenged: "All pigs can fly" is a general statement, but, intuitively, it is not a law of nature, because it is clearly false. Whereas the sentence matches the alleged criterion for lawhood, it is intuitively not a law and thus a counterexample to the proposed analysis of lawhood.

Tests from extensional adequacy presuppose judgments regarding the extension of the concept in question; that is, it presupposes having a strong intuition about which entities or phenomena fall under it or are denoted by it. Preconceptions and intuitions as to what a concept denotes can diverge, however. They may be products of the culture we live in or the way we speak, and professional philosophers' intuitions may well differ from the preconceptions of the folk.

Understanding a concept is not merely a matter of knowing what it denotes. Usually, concepts also carry meanings, or intensions. The so-called Canberra Plan is a complex two-step method for clarifying both the correct extension and intension of concepts. In other words, the Canberra Plan first seeks to fix the meaning of concepts (intension) by describing the role that instances of a given concept have to fulfill then, second, strives to identify its actual fulfillers (extension). It was proposed by philosophers associated with the Research School of Social Sciences in Canberra. First, a concept's use in everyday, scientific, and philosophical contexts is analyzed by collecting all sorts of platitudes about it. A platitude can be anything we say or believe about the concept. For example, regarding causation, we might believe that causes always precede their effects, that nothing causes itself, and so on. By systematizing the platitudes, the Canberra Planners determine which roles the referents of the concept are usually expected to fulfill. In the second step, they then search for referents, that is, entities or phenomena in the world that match these roles. For our example of causation, the transfer of energy could be proposed as such a role player. Because scientific theories are elaborate attempts at describing the world and because Canberra Planners are generally inclined to believe that scientific theories are at least approximately true (that is, they are scientific realists), particular attention is given to the postulates of the sciences. Depending on whether the second step is successful, we may find out the real extension of the concept in question—or we may have to concede that it has no basis in reality and should be discarded. However, note that there are multiple ways of systematizing platitudes and evaluating scientific theories, and hence the outcome may vary.

Apparently, whichever methods we employ, there will always be ways to question our claims in Metaphysics of Science and in philosophy generally. Apart from the proponents of a radical naturalization of metaphysics, philosophers tend to see this not as a fatal flaw but simply as a characteristic feature which is grounded in the very nature of the discipline. The fact that Metaphysics of Science knows no ultimately decisive method but draws on many different tools that may result in different outcomes is not necessarily a bad thing: these tools may just be the best we have to answer questions that we cannot avoid asking, and there may nonetheless be progress in the form of ever more precise, extensionally adequate theories. At the very least, they allow us to map the field of possible views within Metaphysics of Science.

10

JAINA PHILOSOPHY AND METAPHYSICS

10.1 INTRODUCTION

The field of study that surround metaphysics is steeped in knowledge of all things, but is based largely in the abstract. What is identity and what does it mean to "be" or have knowledge? What is the reality of time and space? These are the types of questions that metaphysics deals with, so it shouldn't come as a surprise that questions surrounding various religions are asked as well. Jainism is a religion based in spirituality–a journey from one life to the next as part of an almost endless cycle that culminates when a soul is purified to achieve final liberation from corporeal form. This is a good place to start.

Jaina thought has retained several elements of ancient natural philosophy, such as the theory of atoms, explaining the nature of the universe without recourse to a creator god, and the interpretation of karma as particles of subtle matter.

The Jainas hold an ancient theory in which ultimate reality is divided into two basic categories, one which represents the principle of sentience or consciousness as such, usually translated as 'soul' (jīva), and one which represents the non-sentient principle (ajīva). In their intrinsic natures they are exclusive categories in the sense that, despite their coexistence (as in human beings), the one cannot take on the nature of the other. Both are ontological substances (dravya) in that reality or existence as such can ultimately be reduced to these two eternal, uncreated and indestructible categories.

The soul is best described as a monad, of which there are a countless number, each independent but with the same principle of sentience as its chief characteristic, and each able to expand or contract according to the size of the body which it inhabits. Souls intrinsically possess unlimited qualities of bliss and energy, and are omniscient.

The non-sentient principle is a generic term for the following five substances. First, matter (pudgala) represents the basic stuff or raw material out of which the universe is composed. Matter is made up of atoms, each of which occupies one point of space, and which can be with or without form and can be inhabited by souls. Objects come into being through a combination or aggregation of atoms and change is explained not through destruction or creation of atoms but as a rearrangement of the basic stuff. Together with the view that atoms can have the qualities of colour, taste, touch and smell, Jaina philosophy evolved the theory of substance, quality and mode (see §4) to explain the world in terms of identity and difference, depending on the standpoint and purpose. Matter is crucial for the philosophical anthropology of Jainism because it is matter that directly affects the nature of the soul in the sense that, by becoming converted into karma, it restricts the intrinsic functions of the soul, inhibiting right faith, right knowledge and right conduct. Precisely how this operates is explained later in this section with reference to the so-called 'seven basic truths' which form the basis of Jaina metaphysics.

The second and third substances, called dharma and adharma, are posited for technical reasons. They stand respectively for the factors that assist movement and rest, without themselves being set into motion or coming to rest. They are formless, coextensive with worldly space and without qualities. The function of these two substances is usually explained on the analogy of water and fish. Just as water is not responsible for the fish's movement or its being stationary but makes these possible, so too dharma and adharma offer the condition for the possibility of movement or rest to souls and matter.

The fourth substance, ether or space (ākāśa), does not itself occupy space but is space itself, giving the other substances the place they require in which to exist. According to Jaina cosmology, space is the only substance that encompasses both the world (loka) and the nonworld (aloka). The world is divided into the underworld or hells, in which souls reap the fruits of their unmeritorious deeds, the middle world, which corresponds to our world and in which only human beings can perform the necessary austerities for absolute freedom, the world of different kinds of heaven, in which souls reap the fruits of their meritorious acts and, finally, the world above the heavens, to which a liberated soul (one that has completely rid itself of the burden of all its positive and negative karma) rises, to remain there in its pristine nature. The nonworld serves no purpose at all and is described as being empty. However, it demarcates the area along which liberated souls eternally exist.

Time (kāla) is the fifth non-sentient substance; it has no spatial points, its smallest unit being called a moment (samaya), out of which the other units of time – minutes, hours, and so on – are then derived. Time enables change to take place without itself changing. On the absolute level, time exists as an undifferentiated continuum for the liberated souls. On the cosmological level, it has two broad phases or eras, and is described on the analogy of a wheel with twelve spokes. The world goes through six phases of gradual descent, representing the decline of values, then through six phases of gradual ascent, representing a steady improvement of values. Both phases involve fantastic spans of time. The Śvetāmbara tradition hesitates to accord time the status of an independent substance, though it uses it as a category necessary to explain the course of the world. The Digambaras and the Śvetāmbaras, the two main groups of Jainism, differ in several philosophically minor issues.

What the above ontological structure of reality does not do is explain the original cause of the association of the souls with the other substances, and especially how karma originally attaches itself to the soul to its detriment. This problem applies mutatis mutandis to all schools of Indian thought which speak of the possibility of liberation from the bondage of karma, where the basic issue is simply presupposed and is either left unanswered or euphemistically avoided by saying, as the Jainas also do, that the association between the souls and the world has been so since beginningless time (anādi).

The karma theory was accepted a priori without the need to explain a first cause and the issue does not seem to be regarded as a problem. Moreover, Jainism shares with Buddhism and Hinduism the basic assumption that the human situation is characterized by suffering and pain; more important than seeking the first cause is what can be done to free oneself from this predicament. Past deeds cannot be undone, but once the basic presuppositions are accepted – especially that the process of karma, which is responsible for the cycles of existence, can be interfered with – the individual is free to opt for the path that promises the goal of absolute freedom. The Jainas do not have recourse to a creator god who sets the universe and life in the world in motion. Everything operates through its own inner dynamics and the role of the individual is to meditate on the basic truths of reality.

Jaina metaphysics is based on seven truths or fundamental principles (tattva) which are an attempt to explain the human predicament. The first two are the two ontological categories of the soul and the non-soul discussed above, namely the truth that they exist. For the remaining five truths the factor of karma has to be brought into play because all are connected to it. The third truth is that through the interaction, technically called yoga, between the two substances, soul and non-soul, matter flows into (āsrava) the soul, clings to it, becomes converted into karma and – the fourth truth – acts as a factor of bondage (bandha), restricting the manifestation of the consciousness intrinsic to it. The fifth truth states that a stoppage (samvara) of new karma is possible through asceticism. An intensification of this burns up the existing karma – this sixth truth is expressed by the word nirjarā. The final truth is that when the soul is freed from the influence of karma, it reaches the goal of Jaina teaching, which is liberation (moksa).

It must be noted that the later tradition included two more categories after bondage (bandha), namely punya and pāpa, which may be translated as 'good' and 'bad'. They refer to two kinds of bondage caused by one's deeds. The ethical implications of these terms underscore the significance of Jaina ethics for Jaina soteriology.

10.2 CONCEPTS OF METAPHYSICS IN JAIN PHILOSOPHY

According to Jain thought, the basic constituents of reality are souls (jiva), matter (pudgala), motion (dharma), rest (adharma), space (akasa), and time (kala). Space is understood to be infinite in all directions, but not all of space is inhabitable. A finite region of space, usually described as taking the shape of a standing man with arms akimbo, is the only region of space that can contain anything. This is so because it is the only region of space that is pervaded with dharma, the principle of motion (adharma is not simply the absence of dharma, but rather a principle that causes objects to stop moving). The physical world resides in the narrow part of the middle of inhabitable space. The rest of the inhabitable universe may contain gods or other spirits.

While Jainism is dualistic—that is, matter and souls are thought to be entirely different types of substance—it is frequently said to be atheistic. What is denied is a creator god above all. The universe is eternal, matter and souls being equally uncreated. The universe contains gods who may be worshipped for various reasons, but there is no being outside it exercising control over it. The gods and other superhuman beings are all just as subject to karma and rebirth as human beings are. By their actions, souls accumulate karma, which is understood to be a kind of matter, and that accumulation draws them back into a body after death. Hence, all souls have undergone an infinite number of previous lives, and—with the exception of those who win release from the bondage of karma—will continue to reincarnate, each new life determined by the kind and amount of karma accumulated. Release is achieved by purging the soul of all karma, good and bad.

Every living thing has a soul, so every living thing can be harmed or helped. For purposes of assessing the worth of actions, living things are classified in a hierarchy according to the kinds of senses they have; the more senses a being has, the more ways it can be harmed or helped. Plants, various one-celled animals, and 'elemental' beings (beings made of one of the four elements—earth, air, fire, or water) have only one sense, the sense of touch. Worms and many insects have the senses of touch and taste. Other insects, like ants and lice, have those two senses plus the sense of smell. Flies and bees, along with other higher insects, also have sight. Human beings, along with birds, fish, and most terrestrial animals, have all five senses. This complete set of senses (plus, according to some Jain thinkers, a separate faculty of consciousness) makes all kinds of knowledge available to human beings, including knowledge of the human condition and the need for liberation from rebirth.

Jain metaphysics asserts the existence of principle of sentient or consciousness, called as Jiva or soul, characterised by knowledge and perception. There are infinite, independent souls categorised into: liberated and non-liberated. Infinite knowledge, perception and bliss are the inherent qualities of a soul. These qualities are completely enjoyed uninterrupted by liberated souls, but clouded by karmas in the case of non-liberated souls, resulting in karmic bondage. This bondage further results in a continuous co-habitation of the soul with the body. Thus, an embodied non-liberated soul is found in four territories of existence - heavens, hells, humans and animal world - in an unending cycle of births and deaths, also known as samsara. The soul is in bondage since time immemorial; however, it is possible to attain liberation through rational perception, rational knowledge and rational conduct.

According to Jaina metaphysics, the soul is the master of its own destiny. One of the qualities of the soul is total lordship of its own destiny. The soul alone selects its actions and soul alone reaps its consequences. No god, prophet or angel can intervene in the actions or the destiny of the soul. Moreover, it is the soul alone who makes the required efforts to attain liberation without any godly grace. Amongst the twelve contemplations of Jains, one of them is the loneliness of one's soul and nature of the Universe and reincarnation. Thus, only by purifying one's soul by one's own actions can one help oneself.

Jain metaphysics is also based on seven truths or fundamental principles, also known as tattva, which are an effort to explain the nature and solution of human plight. The first two are the two ontological categories of the soul and the non-soul, namely the axiom that they exist. The third truth is that through interaction, called yoga, between the two substances, soul and non-soul, karmic matter flows into the soul srava, adheres to it, becomes converted into karma and the fourth truth acts as a factor of bondage bandha, restraining the manifestation of the consciousness inherent to it. The fifth truth states that an arrest (samvara) of new karma is possible through austerity, through practice of right conduct, faith and knowledge. An intensification of austerity burns up the existing karma - this sixth truth is conveyed by the word nirjar?. The final truth is that, when the soul is released from the influence of karma, it reaches the goal of Jaina teaching, which is liberation or moksha. Some authors add two additional categories: the meritorious and demeritorious acts, related to karma. These nine categories of primal truth, called navatattva, form the basis of entire Jain metaphysics.

Even though most religions have a reputation for providing faith-based answers to many of life's most profound mysteries, Jainism is important to metaphysics because it does the opposite. While Jainism does provide a number of faith-based teachings and spiritual thought from historical documents left from the ancient world, it also seeks to find answers to the same questions that many scientists struggle with everyday. Because metaphysics deals in the abstract, there are many potential answers to a single question.

Followers of Jainism are the first to acknowledge that the scientific field of physics revolves around what we can prove with empirical fact. What form of matter is this particle? What is its mass? Does it have an observable shape? Physics is broken down into the physical attributes that can be used to describe all matter and how it behaves in relation to other kinds of matter. How is this similar to Jain teachings? You might be surprised.

Jain teachings show that all life is interconnected, and that's why it should be treated with respect. That's why the six Dravyas of Jainism are based in the same physical concepts you would learn about in school.

Dharma is a field of motion. Adharma is rest. Akasha is space. Pudgala is matter. Kala is time. These same ideas and concepts are critical to physics, whereas Jain philosophy deals with the more abstract metaphysical ideas. In Jainism, Pudgala undergoes constant change. A slight suspension of disbelief might allow you to bypass surprise for understanding: Pudgala is broken down into individual units, which are smaller than individual atoms. These smaller units combine into what physics would define as an atom. Jainism takes this one step further by providing other attributes, including color, smell, taste and touch.

Pudgala is also a concept used by Buddhism, but it has a different meaning. When trying to understand the metaphysical world, one could do a lot worse than try to understand the Jain religion and the teachings that provided a great deal of context far before their time.

10.3 EPISTEMOLOGY AND LOGIC

Jainism made its own unique contribution to this mainstream development of philosophy by occupying itself with the basic epistemological issues. According to Jains, knowledge is the essence of the soul. This knowledge is masked by the karmic particles. As the soul obtains knowledge through various means, it does not generate anything new. It only shreds off the knowledge-obscuring karmic particles. According to Jainism, consciousness is a primary attribute of Jīva (soul) and this consciousness manifests itself as darsana (perception) and jnana (knowledge).

Epistemology and logic are closely linked in the Indian tradition because logical issues are largely associated with problems related to inference which, in turn, is regarded as a means or instrument of knowledge. The occupation with abstract, logical ideas related to inference was the favourite topic of the Nyāya school, particularly in its Navya-Nyāya form, which flourished between about the thirteenth and the seventeenth century. Jaina logic adopts much of the Nyāya language and method.

The history of Indian logic can be traced back at least to pre-Christian works on Indian medicine in which physicians are advised to know about a long list of categories, epistemological and logical, when participating in public debates. In the commentary literature to the basic philosophical treatises, especially after the third century AD, epistemology takes a prominent position, each school presenting its own theory of knowledge after criticizing other views. Jainism made its own unique contribution to this mainstream development by also occupying itself with the basic epistemological issues, namely, with those concerning the nature of knowledge, how knowledge is derived, and in what way knowledge can be said to be reliable. These issues feature in the Indian context in the form of the consistent attention paid to what exactly constitutes cognition, the means of cognition, and the validity of cognition. The problems dealt with in epistemology broadly served two functions. They represented an attempt to provide the basis for an intelligible discourse on matters of common, everyday experience and, albeit indirectly, to distinguish this area of discourse from what constitutes the knowledge of ultimate reality. Thus Indian epistemology entails an implicit metaphysical concern when it is employed in the context not only of the knowledge of one's intrinsic nature but also of reality as such.

Knowledge for the Jainas takes place in the soul, which, without the limiting factor of karma, is omniscient. Human beings have partial knowledge – the object of knowledge is known partially and the means of knowledge do not operate to their full capacity. The Jainas have an intricate theory of knowledge concerning the fundamental principles (tattva), and the first systematic presentation of it was given by Umāsvāti in the first chapter of his Tattvārthasūtra (fourth or fifth century). The following is a summary of his theory; the bracketed numbers are the relevant aphorisms. The knowledge of the basic Jaina truths is said to be obtained through means or instruments of knowledge (pramāna) which can yield a comprehensive knowledge of an object, and through particular standpoints which yield a partial knowledge (naya;. The first type of knowledge is of five kinds: sensory knowledge, scriptural knowledge, clairvoyance, telepathy and omniscience, among which again the first two are described as being indirect means of knowledge (paroksa), with the others furnishing direct knowledge, by which it is meant that the object is known directly by the soul (not, for example, as with inferring fire through the cognition of smoke). Jainism is unique in the Indian tradition in regarding sensory perception as an indirect means of knowledge, because it does not take place directly through the soul, although some notable Jaina thinkers (such as Akalanka and Mānikyanandin) are exceptions.

Synonyms for sensory knowledge are remembrance, recognition, 'induction' and 'deduction', all caused by the senses and the mind. Scriptural knowledge, which is based on sensory knowledge, is of two, several or twelve kinds, depending on which scriptural works are taken as authoritative. The range of sensory and scriptural knowledge extends to all the six substances (soul and the five non-sentient substances), but not in all their modes. Clairvoyance is possessed by divine and infernal beings, and can be obtained by human beings and animals through a destruction and/or subsidence of the karma which hinders it. Its scope is all entities that have form. Telepathy is of two kinds, which are distinguished on the basis of purity and infallibility. Its scope is infinitely greater than that of clairvoyance. The scope of omniscience, on the other hand, extends to all substances in all their modes simultaneously. Sensory knowledge, scriptural knowledge and clairvoyance are explicitly referred to as means of knowledge that may also be erroneous.

The basic division of knowledge into the direct and indirect types is generally retained by all Jaina thinkers, though with differences in the way in which they are further classified. For example, Akalanka (c. eighth century) regarded sensory knowledge as a direct means of knowledge, and Mānikyanandin (c. ninth century) evolved a much more complex system with several more subvarieties. When a description of inference as a means of knowledge is given, the occasion is used to discuss logical issues in detail. Jaina logic applies the Nyāya language within its own metaphysical presuppositions. Some of the basic features of Indian logic which the Jainas also adopt involve: (1) the use of inference in the accepted standard form, to convince others of the validity of an argument, clearly distinguishing the three basic terms of the major, the middle and the minor; (2) an example or instance, which can also be in a negative form, based on the invariable concomitance of or inseparable connection between the middle term of an argument and all cases in which the major term obtains; and (3) a concern with types of fallacies based not only on pervasion but also on the relation of the different terms with each other and the locus in which they apply .

With his Nyāyāvatāra (The Descent of Logic), Siddhasena Divākara (c. fifth century) is perhaps the first Jaina thinker who deals with these and other issues in Jaina logic. A few selected examples will show how Jainism imbibed the logic of the Nyāya school, the structure of which it applied to prove its own arguments and to show errors in the arguments of others.

The classical example for an understanding of the abstract ideas expressed here is that of smoke and fire, the basic elements of which, as they feature in a five-step inferential argument, are: (1) the theory – the hill (minor term) possesses fire (major term); (2) the cause or reason – because of smoke (middle term); (3) the example based on invariable concomitance – wherever there is smoke there is fire (as in a kitchen); (4) the application of the theory – similarly the mountain has smoke; and (5) the conclusion – therefore it is the same, that is, has fire.

A statement expressive of the reason (i.e., the mark or the middle term, called hetu) which is inseparably connected with that which is to be proved (i.e., the major term, called sādhya) having been composed of the minor term (called paksa, signifying a side or place), etc., is called an inference for the sake of others (parārthānumāna).

(Nyāyāvatāra, stanza 13; trans. S.C. Vidyabhusan)

Where the inseparable connection of the major term (sādhya) and the middle term (sādhana or hetu) is shown by homogeneousness (sādharmya), the example is called a homogeneous one, on account of the connection (between those terms) being recollected.

(stanza 18)

The reason (i.e., the middle term called hetu) has been defined as that which cannot exist except in connection with the major term (sādhya); the fallacy of the reason (hetvābhāsa) arises from nonconception, doubt or misconception about it (the middle term).

(stanza 22)

The Indians did not develop a system for presenting their abstract ideas symbolically; natural language (pre-eminently Sanskrit), used in such a technical way that only the specialist can decipher the arguments, served the function. However, there have been several successful contemporary attempts to translate Indian logic into the terminology of modern Western logic .

10.4 JIVAS, THE LIVING

As outlined above, the universe is composed of two main kinds of substances, the jīva (living) and the ajīva (non-living). These are un-created existents which are always interacting with each other. These substances behave according to natural laws and the intrinsic nature (sahāvō) of a substance. Understanding this intrinsic nature is the true nature of the Jain dharma.

Jīvas are categorised into two types—liberated and non-liberated. A jīva has various essential qualities: knowledge, consciousness (caitanya), bliss (sukha) and vibrational energy (virya). These qualities are fully enjoyed unhindered by liberated souls, but obscured by karma in the case of non-liberated souls resulting in karmic bondage. This bondage further results in a continuous co-habitation of the soul with the body. Thus, an embodied non-liberated soul is found in four realms of existence—heavens, hells, humans and animal world – in a continuous cycle of births and deaths also known as samsāra. According to Jain thinkers, all living beings (even gods) experience extensive suffering and unquenchable desire (while worldly happiness is fleeting and small in comparison, like a mustard seed next to a mountain). With the exception of the enlightened ones, all living beings are all subject to death and rebirth.

A soul is clothed in various material bodies, of which there are five, each one finer than the other (see image on the right). Every being has at least two bodies, the fiery body and the karmic body. These two bodies don't feel pain or pleasure and can pass through solid matter. A being can have two more other bodies apart from these basic ones, and only the earthly body can be perceived by the eyes. Jains believe that a soul with higher powers can partially leave the body, act outside of it and then return later. This is called samudghata.

According to the Jain philosophy, there are an infinite number of independent jīvas (sentients, living beings, souls) which fill the entire universe. The jīvas are divided into various categories, these include the stationary beings like trees and the beings that move. Jains developed a hierarchy of living beings, depending on the various senses (indriyas) and vital aspects (pranas) that they have. Animals are classed as five sensed being, while plants and various microorganism have one sense. The vitalities or life-principles are ten, namely the five senses, energy, respiration, life-duration, the organ of speech, and the mind. Humans, gods and so on are five sensed beings that also have an inner sense or thinking mind (manas). Regarding sex, the Jains believed that there were three main sexes: male, female and the third sex (napumsaka-veda, all beings without sex organs are part of this third sex). The Jains also affirmed the existence of tiny one-sensed beings called nigodas which exists everywhere and fill the universe.

A unique Jain view is that plants have a form of consciousness like other animals. This is supposed to be seen in their desire for nourishment, reproduction, and self-preservation. They are even seen as capable of expressing moral feelings and thus eventually climbing the ladder of beings towards liberation.

10.5 THE NON-LIVING REALITY

The five unconscious (ajīva) substances (dravya) are:

10.5.1 Pudgala

Pudgala is a term for any non-living particulate matter. The Jains developed an elaborate theory of atomism. Paramāṇus or atoms were the basic and building blocks of matter. They cannot be perceived by the senses and cannot be further divided. An atom also always possesses four qualities, a color (varna), a taste (rasa), a smell (gandha), and a certain kind of palpability such as lightness, heaviness, softness, roughness, etc.

An atom occupies one space point. It is uncreated and indestructible. Atoms combine (bandha) change their modes, and disintegrate (bheda) but their basic qualities remain. An atom can also be bound together with other atoms to create an aggregate (skandha). Material aggregates are categorized according to how fine (suksma) or coarse (sthula) they are. The finest kind of material aggregate is on the atomic scale (extra fine matter), then comes "fine" matter (includes karmic particles), then anything that can be sensed in some way (like smell) but not seen, then comes matter which can be seen but not touched (like light), then there is the category of coarse things (which includes any fluids) and finally there is extra coarse matter (solids). Material things can give off light or darkness. Darkness is seen as a kind of matter in Jainism and so is sound.

10.5.2 Motion/Rest

Dharma and Adharma (Medium of Rest) are substances which account for the principles of motion and rest. As such, they are a kind of aether. Also known as Dharmāstikāya and Adharmāstikāya, they are said to pervade the entire universe. Dharma and Adharma are not motion or rest themselves, but mediate motion and rest in other bodies. Without the medium of motion, motion itself is not possible and vice versa. It is a precondition for movement/rest, like the water which allows for fish to swim. This doctrine is unique to Jainism.

10.5.3 Space

Ākāśa is a substance that accommodates souls, matter, the principle of motion, the principle of rest, and time. It is an all-pervading receptable made up of infinite space-points (pradesha). According to Jains, Space is a substance, in the nature of a vacuum but not a pure vacuum.

It is an extended continuous vacuum. As pure vacuum it will be non-existent, and non-extended; which will devoid it of even one positive quality. Therefore, Jains propound that Space, which is endowed with infinite extension is a substance in itself.

10.5.4 Kala or Time

In Jainism, time is that which mediates change, it causes what is new to become old, and so on. For Jains, time is that which supports the changes to which substances are subject. From one point of view, it is an infinite and endless continuity, from another standpoint, it is made up an infinite number of atomic moments (samaya). Some Jain philosophers hold that time is a substance, while others do not.

According to Champat Rai Jain, "Nothing in nature can exist destitute or devoid of function. Function is discharged by the displacement of energy in the case of simple units and things. If there were no Time-substance to help in the performance of the movement of the displacement of energy, things would be doomed to remain in the same condition always."

Jaina scholars like Gunaratna do not refer to time as a separate substance. He says that time is simply a mode (paryaya) of other substances. But, we need to have a proper explanation of what time or Kala is, and how the Jainas explain it. Kala is an element which cannot be experienced or sensed directly. Therefore, it is not an element of physical experience. Like change and motion which are considered to be real, time is also regarded as real. Unlike the other elements of the universe, which are extended and are astikayas, Kala or time, in its absolutedness, is real. We need to contrast it with 'relative time' and break kala into parts. In reality, kala has no parts. But for all practical purposes, kala or time is divided into parts in order to understand other physical objects in reference to kala or time. The relative time is constituted of simple elements known as KALANUS, or instants. We find that the Jainas, while differentiating kala from the five Astikayas, refer to the Astikayas characteristic of extensive magnitude by a technical term, known as TIRYAK pracaya or horizontal extension. Each surface or extension may be referred to as series. We have arrangements of points or simple elements in a series where one term, as item may belong to another series. This is the two-dimensional series and corresponds to surface or extension. But this is not the case with Time or Kala. Time has only Urdhvapracaya where elements are arranged in forward direction. The series so arranged is mono-dimensional and is said to be in linear order. Therefore, from this, we can say that kala has no extension. It is not an astikaya. But without being an astikaya, it is a distinct and positive real entity, since it brings changes in other things. These are the real characteristics of Real Time or Paramarthika Kala. Thus we contrast Real time with relative or Vyvahara kala. Kala refers to traditional units of long and short periods.

These long and short periods or durations have significance only when they are co-ordinated by a single Real Time series. The essence of Real time is continuity or duration (vartana).Whereas, the essence of empirical time is seen in all changes of the universe. Its is empirical or vyavahara time which we call 'SAMAYA' and which we conventionally and conveniently divide into moments, seconds, hours. We believe that empirical time has a beginning and an end, when in reality; it is beginning less and endless. Kala is the only non-extended substance, while the rest are astikayas or extended. Unlike Relative or Vyvahara time, Real or Paramarthika Time is indivisible, and is formless. It is eternal. Kala is Anastakaya.

Kala or time, though not an element of the physical universe, may be mentioned here. Since change and motion are admitted to be real, time also must be considered real. The real or absolute time as contrasted with the relative time is constituted by simple element known as kalanus or instants. In stance points and atoms are the characteristic conceptions of Jaina thought and in this respect it has a wonderful corroboration from the field of modern mathematics.

The Jaina thinkers in distinguishing time from the five astikayas made use of an important idea. Astikaya is spatiality or extensive magnitude. This extensive magnitude is denoted by a technical name. Tiryak ,pracaya or horizontal extension.

When the simple elements, say, the points are so arranged in a series where each term is an item also in another series we must have the two-dimensional series which will correspond to surface or extension. Wherever there is such a tiryak-pracaya we have astikaya. But time or kala has only urdhva-pracaya. The elements are in a forward direction. The series is mono -dimensional or linear order. Therefore kala has no extension either directly or indirectly. Hence it is not an astikaya. Though it is not an astikaya it is distinctly real entity which accounts for changes in other things.

Such are characteristics of real time. This should not be confounded with Vyavaharra kala or relative time which is measured by some conventional units of either long or shorts duration. These conventional distinctions would have no meaning if they are not co-ordinated in single real time series.

10.6 THE SIX DRAVYAS

The term dravya denotes any existence which has the important characteristic of persistence through change. Jaina conception of reality excludes both a permanent and unchanging real of the permenidian type and also the mere eternal flux of Haraclites. An unchanging permanent and mere change without substratum are unreal or impossible abstractions. Jaina system admits only the dynamic reality or dravaya. Dravya then is that which has a permanent substantiality which manifests through change of appearing and disappearing. Utpada- origin, vyaya-decay and dhrauvya--- permanency form the triple nature of the Real. To emphasizes the underlying identity alone would end in Vedanitc conception of this Real as Brahman. To exphasise the change alone would result in the ksanika- vada of the Buddhist, the reality as a stream of discrete and momentary elements. The Concept of darvya reconciles both these aspects and combines them into an organic unity. It is an identity expressing through difference, a permanency contenting through change. It corresponds to the modern conception of organic development rather in its Hegelian aspect. It has duration; it is movement; it is the Elan Vital. The five asitikayas and kala or time six dravyas or the real existences.

10.7 DRAVYA, GUNA AND PARYAYA OR SUBSTANCE, QUALITIES AND MODES

Having introduced the Jaina concept of the real let us examin the distinctions in the realm of the real. The dynamic substance or dravya is always associated with certain intrinsic and inalienable qualities called gunas. Thus the yellow colour, malleability , etc; will be the qualities or guna of the dravya gold. The Dravyas with its inalienable qualities must exist in some state or form. This is its mode of existence or paryaya. This mode or paryaya is subject to change. It may be destroyed and a new mode may appear. But this creation and destruction are relevant only to paryayas or modes and not to Dravyas, the constitutive substance. That can neither be destroyed nor created. That is eternal.

The approximate parallel conceptions in the western thought will be Spinoz`s substance, qualities and modes. Of course, for Spinoza, there was only one substance whereas here we have six distinct substances or dravyas . The term attribute is used in a technical sense by Spinoza whereas it means merely the qualities in Jaina Metaphysics. Each dravya or the real has its own appropriate. Matter has the sense qualities of touch, taste, etc; soul has cetana consciousness; and so with the others dravyas. What is the relation between dravya and guna the reals and their qualities? This is an important problem even in modern metaphysics. There have been two distinct answers proposed by philosophers One answers always emphases the permanent basis as the real and the other as the changing qualities. The former condemns change as mere appearance or maya, whereas the latter condemns the permanent substance as a mere figment of imagination. The former school generally ends in agnosticism and maintains that the real is unknown and unknowable. The latter generally ends in scepticism and sweeps always in a flood of doubt the fundamental concepts of life and world, of morality and religion. In India we have Advaita Vedanta as an example of the former and Buddhism, of the latter. In the west the philosophers like Locke and kant may be mentioned as representatives of the former school and Hume may be taken as a type of the latter.

The problem once again taken by Bradley who after an acute analysis condemns it to be insoluble. Hences according to his own dialectic the thing with its qualities is self-contradictory and therefore an appearance .this is a conclusion which goes clean against science and common sense. Thing-hood is not after all a self-contradictory notion. For the Jainas it is a fundamental concept. The thing or Dravya as it is a dynamic entity is always flowing. It has no existence apart from its qualities which in their turn are not really distinct from the dravya. The thing-in itself apart from all the qualities is merely an empty abstraction. The qualities themselves are not merely monentatry fleeting sensations. The thing exists in and through its qualities and the qualities related and organised constitute the thing. The difference between the two is only a difference of reference and not a difference of existence; in the technical language of Jaina metaphysics, the anyatva between dravya and guna is only vyapade- matra. They do not have pradesa prthaktva. Hence they are one in reality though having distinct nomenclature and reference.

Paryaya is another technical term demanding careful understanding. Paryay means mode of existence. This again is viewed from two different aspects, artha-paryaya and vyanjana- paryaya.we have already mentioned that dravya is but an entity that is continually changing. This triple nature of reality that is permanency through births and deaths, through creation and destruction gives to dravya a characteristic mode of existence every moment. This cotinuous flow of the real is parallel to the continuous flow of the duration of time. This intrinsic change of dravya is known as artha-paryay. All the six dravyas have this artha-paryaya . what is vyanajana paryaya has a pretty fixed duration of existence. Besides the molecular aggregation and disintegration that take place every moment in a physical object the object may have a particular mode of existence as a pot, for example for a certain duration of time. This paryaya of pot is vyanjana- paryaya of pudgala. Similarly for Jiva. The continuous change that takes place in consciousness is Jivas artha- paryaya. Its existence as a particular organism as a man or a deva with determinate age is the vyanjana-paryaya of Jivas and pudgala have got only artha-paryayas alone. The reals are thus exhausted by the six dravyas with their respective gunas and paryayas.

Since these dravyas are reals they have satta or astitva or existence as there common characteristic. From this point of view of satta, all the deavyas may be brought under one class. Though from the class-point of view all the dravyas are one. Still the satta, their common characteristic should bot be abstracted and postulated as the unitary substance of which the other dravyas may be taken as paryayas. ekanta view is condemned as unwarranted and erroneous. The six dravyas, in spite of their common characteristic of satta, are fundamental and irreducible from one to another.

One more point and we may leave this topic. A thing in concrete world is therefore a paryaya of anyone of the dravyas. It is also otherwise called artha. An artha. Or a thing is a corporate unity oaf an infinite number of qualities, just as the cosmos is a system of infinite number of arthas. The one is extensive and the other is intensive; but both are infinite wholes: According to Jaina philosophy therefore, we require an infinite thought to apprehend them completely. He who cannot know a thing completely cannot know the world completely and conversely he who cannot know the world completely cannot know even a single thing really and completely. This particular attitude of the Jania thinkers reminds us of Tennyson`s lines in, `A flower in the crannied wall I I could know the root and all I could know what God and man is."

Having spoken of the dravyas general , let us try to explain Jiva-dravya a little more in detail.

10.8 JAINA BIOLOGY

Perhaps it is inaccurate to speak of biology in the system before us. The science of biology as such is peculiar to modern age; hence we are not quite justified in expecting such a scientific conception in a work of pre-Christian era and which is perhaps of the same age as of Plato and Aristotle. Naturally therefore, the ideas about the organic world are curiously intermixed with various mythic and fantastic conception relating to being of hell and savage or heaven. Eliminating all these as irrelevant we still have considerable material to enable us to have an insight into the ancient ideas about life and living beings.

One important conception that would strike the reader in the very beginning is the organic unity of the plant and the animal world. Vegetable kingdom is distinctly organic. Its nature was accurately observed and carefully described. The whole plant world is included in the class of organisms having one sense, i.e. the sense of contact. Then the world of animals and insects is classified according to the same principle of sense organs. The animal world beginning with such insects as earthworm and ending with man is brought under four main group, organisms, with two three four and five senses.

10.9 PRANAS

Associated with organisms there is the concept of pranas or the essential characteristics of living beings . these are mainly four, bala- prana, indriya ayuh- prana. Every organism implies certain capacity of spontaneous activity. This capacity for action is bala-prana or life potency. Every organism must possess some kind of sense awareness. This implies the possession of a sense organ and the capacity to apprehend the environment through that sense. The number of sense organs is different according to stages of organic development. Next is ayuh-prana or duration of life. Every living organism has a limited duration of life. This organic capacity to persist through a certain duration is ayuh-prana. And lastly there comes respiration. There is no organism without this prana of ucchavasa- nihsvasa. These four main pranas are the essential attributes of organic beings.

10.10 DIFFERENT KINDS OF BIRTHS

One other interesting point is the enumeration of the different means of birth of organisms. Young ones may be produced from garbha. These are garbhajas, young ones produced from the womb. Then the andajas. The young one produced from egg. Thirdly there is sammurchana or spontaneous generation. This refers to minute organisms. And lastly, they speak of aupapadikasi. I the case of devas and narakas. The last one of course we may treat as beyond scientific pale. The recognition of spontaneous generation is a point deserving special emphasis.

10.10.1 Microscopic Beings

Another fact deserving notice is the early recognition of the existence of microscopic organisms. These are called suksma ekendriya jivas or microscopic organisms having only one sense. These are said to fill the earth, air water and fire. The possibility of microscopic organisms in fire seems a little too fantastic.(for the verification of this one must look to the result of further scientific investigation) But in the case of the other three we need have no hesitation as they are fully established by modern science.

In this connection it is necessary to point out that H. Jacob's conjecture that Jainism is very ancient though historically true rests on an unwarranted assumption as to this jiva-nikaya. He interprets these jivas in earth, air water and fire in such a way as to suggest that Jainism is or was once at least, a mind of Indian fetishism which believed in the souls of earth, air, etc. Then what is purpose of calling these suksma ekendriya Jivas? What is the meaning of this distinction between Jiva and ajiva, cetana and acetana?

The different organisms mentioned above are jivas or souls in association with matter or pudgala. If you look at the system of metaphysics as a whole you have the picture of an infinite number of jivas caught in the vortex of matter: surrounded by a soulless environment. This picture is extremely analogous to the cosmic picture of the Sankhyas. An infinite number of Purusas submerged in an ocean of prakrti and drifting along the current of prakrtic evolution. But here instead of one homogeneous prakrti constituting the environment of the purusas we have the five ajiva-dravyas forming the appropriate stage for the enactment of the spiritual drama of souls. The primary characters are jivai and pudgala. The other make the things go. The cosmic evolution is due to the interaction between soul and matters as samsara. How could acetana pudgala bring about such a calamitous result? Pudgala or matter in a very subtle and fine form fills the cosmic space. This form of pudgala is spoken of as karma-prayogya-pudgala matter fit to manifest as karma.

In this cosmic space are also the jivas. Jivas by their impurity of heart led to formation of subtle material cocoon of karmic bodies which retard and obstruct the intrinsic spiritual radiance of the soul. When once this subtle deposit of matter is begun the grosser encrustation of matter and the formation of organic bodies of matter is begun. The grosser encrustation of matter and the formation of organic bodies is inevitable according to Psycho-physical laws. Here we have to notice one important point. Even such great oriental scholar like Hermann G. Jacobi makes the ambiguous to and misleading statement that karma according to Jainas is purely materials and the suggests that since they speak of a combination between jiva and karm, Jainas assume that soul is of some subtle matter thus making the combination between the two possible. This is extremely misleading and far from the truth. There is neither combination nor direct causal relation between soul and matter. Jainas speak of bhava -karmas as distinct from dravya-karmasi. The former psychological karma is immediate to jiva.

The later material is mainly concerned with the building up of the bodies. No doubt the two are associated together; yet the two are distinct and separate. Jiva is cetana and sarira is acetana.

This conjoint existence or samsara is anadi without beginning. Through its congenital impurity of the heart soul is thrown into impure environments from time immemorial, though there can be an end to the state of existence. This end consists in securing spiritual freedom and perfection through obtaining the purity of the heart which in its turn leads to the destruction of the material encasement. Thus the life of the soul in samsara an infinite series without beginning but with an end which is moksa.

10.11 SOUL AND BODY

Every organism or a samsari-jiva is an organic unity of two distinct entities, jiva and pudgala, soul and body. Naturally therefore there crops up the problem of the relation between the two. Soul is cetana (consciousness,) amurta (no- corporeal,) arupa (non-sentient) whereas body has the opposite qualities in each case. One may be said to be the contradiction of other. The dualism is so emphatically expressed here as in Cartesianism.

The term, body, implies two different things. The gross body that we actually perceive through our senses. This is constituted and nourished by matter taken in the form of food, etc this body is every moment changing and will be given up by the soul after a certain period. Besides this gross body there is for every jiva a subtle body known as karmana sarria. This body is constituted by subtle material molecules known as karama-pudgala. This is transcended only in the perfect state. In discussing the relation between the states of this karma-sarira and the state of jiva Jainism makes an important distinction between upadana-karta and nimitta-karta, substantial cause and external cause. Mental states are the modifications of the mind and physical states are the modifications of matter. Mind is the upadana-karta of psychical states and matter is the upadana-karta of physical changes and yet physical states and psychical states may be mutually external conditions. The causal activity contemplated here is a bit obscurer. One psychical state is due to the immediately antecedent psychical state and similarly one physical state and one physical state is the result of its own antecedent. Thus mental series in a way is independent of physical series. But still mental change may be externally determined by a physical change and the physical and the mental is purely external. In the technical language of the system one is the nimitta-karta of the other.

So far as we are able to make out, the meaning seems to be this: a mental change is due to two conditions, one an upadana- karta, a mental antecedent and nimitta-karta a physical antecedent. The mental change is the result of both these antecedent conditions, physical and mental. Similarly a change in the body is to be traced to two conditions: an upadana condition, a physical antecedent in this case and a nimitta condition, a mental antecedent.

The system emphasises the causal interrelation between mind and matter, even though the interrelation is one of external condition. The reason given for accepting this interrelation is the reality of moral responsibility. If there is no causal interrelation mind and matter why should a person taken responsible for his conduct? If moral responsibility is real, if moral evaluation of conduct is genuine. Then conduct must be the intimate expression of the personality.

Though the discussion is between jiva and its karmana sarira,the discussion and its conclusion may very well be taken as relevant to our problem of the relation between soul and body . The whole discussion may be taken as expressing the views in regard to the wider problem. Soul and body are capable of causal interrelation and a change in one always involves two Antecedents, one physical and the other psychical. If causal interrelation is not admitted certainly ethical value will remain unexplained and unintelligible.

10.12 SENSATIONS AND SENSE ORGANS

The sense organs recognised in the system are the usual five. But sometimes manas or mind is also spoken of as an indriya. Indriys in general are of two kinds, dravya indriya or the physical sense organ and bhava indriya the Psychical counterpart. Sensory awareness dravya indriya and the physical object. It assumed of course that only physical objects or pudgala that can be apprehended by sensation.

This contact may be direct or indirect. In the case of sight the contact is indirect. The object perceived by vision is not brought in contact with the eyes. The objects in space are revealed to us by light or jyoti. It is through being illuminated they are apprehended by vision. The exact operation of light on the eyes is not further explained. In the case of the other senses we have direct contact. But the direct contact may be sthula or suksma, gross or subtle. In the case of contact and taste we have the direct contact with the gross object.

But in the case of smell we have contact with minute particles of the object smelt. In the case of sound also we have suksma contact. But in this case what the ears come in contact with is merely a kind of motion Unlike the other Indian systems of thought which associate sound with akasa, Jaina system explains the sound as due to the violent contact of one physical object with the other. It is said to be generated by one skandha kocking against another skandha. Sound is the agitation set up by this knock. It is on account of this theory of sound the system speaks of an atom or parmanu as unsoundingi by itself. Thus in all these cases the environmental stimulus is either directly or indirectly a physical object. Sense perception is the result of the contact between two physical things, dravyendriya on the one hand and the stimulus from the object on the other hand .

10.13 ANALYSIS OF SENSATIONS

The next interesting point is the analysis of the different sensations obtained through different sense organs. Through the eyes we have the apprehension of five colours Visual sensations consist of the five elements or panca-varna. But we have to note here that sensation of white is also included as one of the colors. In this respect the term svarna or colour is used in its popular sense and not in the scientific sense. Similarly taste is of five kinds, pungent, bitter , sweet, sour and saline.

These five tastes are obtained through the tongue, which is rasanedriya. Skin is sparsanendriya and through it the following eight kinds of coetaneous sensations are obtained: light and heavy, soft and hard, rough and smooth and cold and hot sensations, four pairs of opposite senses. These coetaneous sensations include sensations of temperature, contact pressure and muscular or kinesthetic sensations. Sensation of smell is only of two kinds, sugandha and durganadha. Sound sensations are of infinite variety. The different kinds of sounds natural and artificial, purposive, and non- purposive, articulate and inarticulate, musical and non-musical are spoken of.

10.14 SENSE PERCEPTION

What we directly apprehend through a sense organ is to not merely particular sensation but the object the object. Sense perception is known as darsana. Darsana is the perception of a physical object. Darsana may be caksur-darsana and acaksur-darsna means perception of an object through visual sensation. Acaksur- darsana. Means perceptian through the other senses. Darsna or sense perception not only implies the passive receptivity of the mind but also the active interpretation of the received stimulus, i.e darsana means the com placation between the datum and metal construction.

This is implied in the description given of `knowledge by acquaintance` or mati, avagraha iha, avaya, are different stages of sense perception, avagraha refers to roughly the datum. But the datum does not mean anything, it is merely the understood patch of colour, e.g.; in the case of visual sensations. At the presentation of this visual patch there is the questioning attitude of mind which is represented by the term iha. As a result of this examination we may interpretation is avaya.

In the case of visual perception these three different stages may not be clearly distinguishable. But in the case of auditory perception we may clearly recognise the different stages. Darsana then includes all these three stages, then only is the thing known to us.

In this connection we have to notice one important point. The term darsana is not confined to sense perception. It is a general term including the sense perception as well as the supernormal perception of other kinds. Two kinds of supernormal perception as well as the supernormal perception of other kinds. Two kinds of supernormal perception are general mentioned by Jaina thinker acadhi-darsana and kecala-darsana. Avadhi-darsana refers to the peculiar king of clairvoyant capacity which is able to perceive things and events in distant places and also in distant times either past or future. Objects and events not evident to the normal sense perception are obvious to Avdadhi-darsana. But the object of avadhi perception appear as if they are perceived normally close at had. If is said that Avadhi-darsana is concerned with only rupi-dravyas or perceptual object. The other darsana known as kevala-darsana is perception par excellence. It is associated with perfect consciousness. This faculty is acquired only after complete emancipation from karmic bondage. To this perfect perception the whole reality is obvious. In short it refers to the all-perceiving faculty of Paramantma. What we are justified in speaking of in connection with Jaina psychology are the normal sense perception (caksur-darsana and acaksur- darsana) and the supernormal clairvoyant perception (or avadhi-darsana).

10.15 JNANA OR KNOWLEDGE

Jaina account of cognition's also interesting. Jana or understanding is said to be of different kinds according to means employed in cognition.(1) ,mati-Jnana, is knowledge obtained through the normal means of sense perception and memory based upon the same. This is the common inheritance of all persons (2) Srita-Jnana is knowledge obtained through testimony of books. This corresponds to knowledge by only the learned men. Besides these two means of knowledge there are three other supernormal means of understanding. These are avadhi-Jnan, manahparyaya-jnana. and kevala-jnana.

Avdhi-jnana is the understanding of the nature the objects obvious to avadhi-Darsana. Manaparyaya- jnana refers to a peculiar kind of telepathic knowledge acquired by which knowledge of alien minds is obtained. The last one of course refers to the perfect understanding or the omniscience of the Perfect Being Or purusottma. Treating this as the metaphysical ideal we have to recognise the other four kinds of cognition as relevant to our psychological interest

10.16 AFFECTION

Affective consciousness plays a very important part in Jaina metaphysics. The whole religious discipline is directly secured by a stoic freedom from the affective influence of environmental objects. Experience of pleasure pain ,is assumed to be the specific characteristic of organised beings or samsari jivas it is mentioned that jiva has the tendency to continue beneficial activity from which pleasure results and to discontinue harmful activity from which pain results. This is so very analogous to biological description of the instinct of self-preservation. Jiva equipped with this quality naturally desires pleasant things and avoids unpleasant things.

Since the psychological analysis is subordinate to the metaphysical system several facts of psychological interest are throw into the background of the philosophical scheme. Nevertheless there is no mistake about the striking psychological analysis exhibited by Jaina thinker. Experience of pleasure and pain, is generally referred to as karmaphalacetanar or consciousness of the fruits of action. Pleasure and pain are always viewed in relation to action.

Bhava or affective consciousness is of three kinds, subha-bhava, asubha-bhava, and suddhai-bhava. Feeling of pleasant nature feeling of an unpleasant nature ,and feeling of pure nature. The last one refers to the enjoyment of Self by Self. As such it may be taken to mean the spiritual experience of the pure Self. The other two kinds of the feeling are relevant to the point. These are corresponding to the normal feelings manifest as the result of karma or action and on the other hand they are determined by objects in the environment.

A very interesting classification of emotions is given in connection with the conditions of karmic bondage. These emotions are generally decided two main-classes sa-kasaya and a-kasaya, those that have the tendency to colour or stain the purity of the soul ad those that do not have that tendency. The sa-kasaya ones are krodha or anger, mana or pride, maya or deceitfulness or dissembling and lobha or greed. The emotions are : hasya-laughter; rati- feeling of attraction; arati- feeling of repulsion; soka sorrow; bhaya fear jugupsa feeling of disgust which may manifest in hiding once own weakness; striveda –peculiar sex feeling of women pumveda- peculiar sex feeling of men; ad napumsaka-veda- the corrupt sex feeling of eunuchs.

Again certain instinctive tendencies are also referred to as samjnas. These are ahara, bhaya maithuna ad parigraha hunger fear sexual appetite ad acquisitive instincts. These are corresponding feelings to the instinctive appetites which may colour the consciousness of a jiva.

The feelings aspect of sensations is implied in the very classification of the sense elements. The feelings aspect is predominant in the case of smell and taste whereas it is indirectly associated with auditory and visual sensations. The rest of the references to feeling of pleasure and pain are purely metaphysical and therefore they are more of religious interest that of scientific interest.

10.17 CONATION OR THE CONSCIOUSNESS OF ACTION

Atman is not only jnani and bhokta, the knower and the enjoyer, but is also a karata or the agent. This may be considered as the central idea of Jaina system. Soul by its own activity is able to make or mar its own destiny. The Theory of karma is intimately associate with the causal agency of Atman. As the result of this metaphysical assumption we have several facts of psychological importance mentioned in the system. Even in the lowest organism there is the tendency to continue pleasurable activity and to discontinue painful activity. This primitive tendency of life or jiva is just the cognitive activity which develops into conscious choice of an end or purpose which is the characteristic of volitional activity. In human beings this cognitive tendency is naturally associated with raga and dvesa, desire and averision.

Cognitive activity in general is denoted by the term karma-cetana. This karma-cetana or consciousness of activity is to be associated with the zoological kingdom, trasa jivas. The plant world or the world of ekendriya sthacara jivas is devoid of this karma-cetana. They have karma phala –cetana alone whereas the other jivas have both and also jnana cetana to boot. The importance of volitional activity is clearly testified by the part it plays in the Jaina system of ethics. The psychology of will is also connected with another doctrine of psychology of will is also connected with another doctrine of psychological importance. Mohaiya karma which is considered to be the root of all evil has two aspects cognitive and connective. What is known as drasana mohaniya interferes with the faculty of perception and belief. Caritra mohaiya is a sort of corruption of the will; it misleads the will and thus leads the jiva towards evil. We shall consider the relation between karma and atman when we go to consider the ethical aspect of Jaina system. In the meanwhile let us see what Jaina logic is.

10.18 ETHICS

Given that the proper goal for a Jain is release from death and rebirth, and rebirth is caused by the accumulation of karma, all Jain ethics aims at purging karma that has been accumulated, and ceasing to accumulate new karma. Like Buddhists and Hindus, Jains believe that good karma leads to better circumstances in the next life, and bad karma to worse. However, since they conceive karma to be a material substance that draws the soul back into the body, all karma, both good and bad, leads to rebirth in the body. No karma can help a person achieve liberation from rebirth. Karma comes in different kinds, according to the kind of actions and intentions that attract it. In particular, it comes from four basic sources: (1) attachment to worldly things, (2) the passions, such as anger, greed, fear, pride, etc., (3) sensual enjoyment, and (4) ignorance, or false belief. Only the first three have a directly ethical or moral upshot, since ignorance is cured by knowledge, not by moral action.

The moral life, then, is in part the life devoted to breaking attachments to the world, including attachments to sensual enjoyment. Hence, the moral ideal in Jainism is an ascetic ideal. Monks (who, as in Buddhism, live by stricter rules than laymen) are constrained by five cardinal rules, the "five vows": (1) ahimsa, frequently translated "non-violence," or "non-harming," satya, or truthfulness, asteya, not taking anything that is not given, brahmacharya, chastity, and aparigraha, detachment. This list differs from the rules binding on Buddhists only in that Buddhism requires abstention from intoxicants, and has no separate rule against attachment to the things of the world. The cardinal rule of interaction with other jivas is the rule of ahimsa. This is because harming other jivas is caused by either passions like anger, or ignorance of their nature as living beings. Consequently, Jains are required to be vegetarians. According to the earliest Jain documents, plants both are and contain living beings, although one-sensed beings, so even a vegetarian life does harm. This is why the ideal way to end one's life, for a Jain, is to sit motionless and starve to death. Mahavira himself, and other great Jain saints, are said to have died this way. That is the only way to be sure you are doing no harm to any living being.

While it may seem that this code of behavior is not really moral, since it is aimed at a specific reward for the agent—and is therefore entirely self-interested—it should be noted that the same can be said of any religion-based moral code. Furthermore, like the Hindus and Buddhists, Jains believe that the only reason that personal advantage accrues to moral behavior is that the very structure of the universe, in the form of the law of karma, makes it so.

11

IMPACT OF COVID-19 ON CHILDREN RIGHTS

11.1 INTRODUCTION: AN OVERVIEW

> *This is an unprecedented crisis and it presents unprecedented risks to the rights and safety and development of the world's children. Those risks can only be mitigated through unprecedented international solidarity for children and humanity.*

- Writing over a year since the COVID-19 pandemic started in 2020, a narrative is well-rehearsed. While responses to the pandemic show the potential for the impossible to become possible, they have also highlighted long-standing issues and exacerbated a range of troubling inequalities.[1] This narrative is as applicable for children and children's human rights, as it is for other groups and issues.
- In the rush to protect the human right of survival and development for children, families and others in their communities, the emergency policies[2] and their implementation magnified the challenges in taking a children's rights approach in adult-oriented systems and institutions. This article explores such challenges, drawing on policy analysis undertaken in Scotland. The Children and Young People's Commissioner, Scotland (henceforth, the Commissioner) asked the Observatory of Children's Human Rights Scotland to undertake an independent Children's Rights Impact Assessment (henceforth, independent CRIA) on the COVID19- policies that most affected children living in Scotland. This was an intensive exercise undertaken from 11 May until 19 June 2020, intended to unveil and inform the accumulating concerns that children's human rights were being violated in the urgent response to the pandemic.
- It goes on to consider the repercussions of constructing children as 'vulnerable' as well as the vulnerability of children's rights in policy making and concludes by questioning the application of children as vulnerable, and as particularly vulnerable. Going forward, to ensure compliance with children's human rights,

the article recommends greater emphasis on children's rights accountability to mitigate and improve policies, both in relation to pandemic responses and future incorporation of rights into domestic law.

11.2 CHILDREN'S RIGHTS IMPACT ASSESSMENTS

- Children's Rights Impact Assessments (CRIAs) are a means to analyse and predict the impact legislation, policy or budgetary allocations have on children and the fulfilment of their rights. The European Network of Ombudspersons for Children provide the following definition:
- A Children's Rights Impact Assessment examines the potential impacts on children and young people of laws, policies, budget decisions, programmes and services as they are being developed and, if necessary, suggests ways to avoid or mitigate any negative impacts.
- CRIAs apply a human rights analysis to a particular group – children and young people – and thus sit within the suite of human rights impact assessments. As discussed by McCall-Smith, their value arguably lies in moving away from the performativity of presenting activities, to concentrating on the actual or potential impacts of the ensuing practices on people's human rights. Further, they can identify negative impacts (and how to mitigate them) and identify further positive impacts (and how to amplify them). The independent CRIA sought to evaluate the actual and likely impacts on children's human rights of the COVID-19 policies. It further aimed to look ahead: what could be learnt should there be another lockdown? for services more generally post-pandemic? and for the CRIA process itself?
- The independent CRIA was unusual, in terms of covering a number of policies rather than a single one, responding to policies that were already in place, the impacts of which were unfolding,and being undertaken by experts rather than officials in public bodies. Policies were selected because they were introduced by the UK or Scottish Governments in response to the COVID-19 pandemic and impacted children in Scotland. An initial screening was conducted by the Commissioner's Office to identify the core policies. Nine teams of academic and other experts undertook intensive analysis of these policies, in light of human rights obligations and available evidence, on key subject areas. The teams followed a CRIA template, adapted from the Scottish Government's Children's Rights and Wellbeing Impact Assessment. An overview report was then compiled, that drew in key findings from these subject areas as well as addressing other key human rights issues arising. To support the process throughout, a database of evidence was collated, and additional specialists were asked to contribute and review before, during and in the final stages of the independent CRIA. A particular effort was

made to include evidence from children and young people, of which there were several quantitative and qualitative sources, as well as evidence from the Children's Parliament and the Scottish Youth Parliament who are founding members of the Observatory of Children's Human Rights Scotland.The Commissioner's Young Advisers considered the independent CRIA's findings and produced their own recommendations; the Children's Parliament worked with children to ensure a child-friendly version of the independent CRIA was published.

- The independent CRIA was undertaken in favourable conditions for children's human rights. Scotland is emerging as a world leader in promoting human rights generally and children's human rights in particular. The United Kingdom, as the State Party, ratified the UN Convention on the Rights of the Child in 1991 and has thus been obligated for over 30 years to operationalise the UNCRC in practice and policy. Scotland has long had separate legislation and legal jurisdiction in areas such as education, social services, health, family and criminal law, and this was only extended since Scottish devolution and the (re)establishment of the Scottish Parliament in 1999. The Scottish Government has increasingly articulated a commitment to human rights generally, with the establishment of the First Minister's Advisory Group on Human Rights Leadership and to children's human rights specifically. For nearly two decades, the Scottish Government has used variations of the slogan 'making children's rights real' as an organising principle of its childhood policy. Thus, children's human rights should have been a central feature of policy-making under COVID-19.

- However, the independent CRIA found children's human rights were far from a central feature. Fundamentally, the policies introduced to respond to the COVID-19 pandemic sought to protect the health of children and their families, thus respecting their human rights of health, survival and development under the UNCRC and other international human rights treaties. Government was under unprecedented pressure to develop new policies to respond to the pandemic and particularly the first 'lockdown' from 23 March to 29 May 2020, where virtually all people were required to remain indoors, to minimise face-to-face social contact, and to have restricted access to public spaces. The Scottish Government took steps to consider children's rights: for example, it undertook and published three children's rights and wellbeing impact assessments on three key COVID-19 policies and mapped out its responses according to the 11 recommendations of the UN Committee on the Rights of the Child.

- It established working groups to regularly review and plan for policy and practice responses, for childhood-related services, and published updated information and actions.Efforts, for example, were made to prevent digital exclusion of children, particularly in relation to home-school learning, to set up systems that could address child protection and domestic abuse, and to ensure basic needs for food were met. Nonetheless, the independent CRIA highlighted key areas that raised significant concerns for children's human rights in Scotland, such as:

- Detrimental short-, medium- and long-term impacts on children and young people's physical, mental and oral health, on their levels of physical activity and their education.
- Harmful restrictions, delays, and suspension of specific legal rights for children and young people in such areas as young offenders' institutions, criminal justice more generally and child protection, additional support needs in education, young carers' assessments, state care, acute mental health needs, birth registration and civil justice.
- The shift to digital communication and learning excluded some children and young people who had disadvantaged access and/or who had particular communication needs.
- Certain children and young people risked being even more disadvantaged, across areas from health to play to education, including those with disabilities, with additional support needs, who were carers, who were from Gypsy/Traveller communities, who were from black, Asian and minority ethnic communities, who were from socio-economically disadvantaged households, and who were refugees or asylum seekers.
- Restrictions of movement risked criminalising young people aged 16 to 17 initially, increased vulnerability for children and young people who were not safe in their households, and prevented children and young people having physical contact with friends and family members who were not living in their households.
- Worsening economic inequalities, which were not adequately dealt with by social security changes nor alternatives to free school meals.
- Face-to-face educational and early years provision were restricted to children of key workers or who were 'vulnerable', which initially had poor take up due to stigma, limited transport, and other reasons.
- The lack of information for children and young people, the lack of engagement with children and young people in policy decision-making, and the lack of disaggregated quantitative data on children and young people.
- The independent CRIA goes far further, with an extensive and specific list of gaps and problems for children's human rights as a result of the COVID-19 policies. In a context where there was such official government and policy commitment to children's human rights, why were there so many gaps and violations of them?

11.3 CHILDREN AND CHILDHOODS IN DISASTERS

- Childhood studies provide key conceptual ideas that can help us understand the gaps and violations for children's human rights. As developed particularly in the 1990s in the Global North, childhood studies put forward that childhood was a social category that was worth considering in its own right. And, as with other social categories like race, gender or disability, childhood could be understood as a social construction and socially constructed. This was effectively explained by Prout and James:
    - A child's immaturity is a biological fact: but how this immaturity is understood and how it is made meaningful is a cultural one.
    - The dominant constructions of childhood, in the Global North, depicted children as vulnerable, dependent, innocent and incompetent. Rather than recognising children's expressions of agency, children were valued as future capital and human investments: as 'human becomings' rather than 'human beings', as Qvortrup famously wrote.
    - In response, James and Prout suggested a new paradigm for understanding and researching childhood. Several points are particularly useful for our consideration here:
    - Childhood is a variable of social analysis
    - Children's social relationships and cultures are worthy of study in their own right
    - Children are and must be seen as active in the construction and determination of their own social lives, the lives of those around them, and of the society in which they live
    - Proclaiming a new paradigm of childhood is also part of reconstructing childhood in society
- The social constructions of childhood have profound implications for policy. If children are predominantly perceived as vulnerable and dependent, then protectionist strategies are privileged that emphasise parental responsibilities and child safety. If children are predominantly considered 'empty vessels' and future citizens, then educational and related social training are emphasised to ensure children grow up to be contributing and responsible adults. In contrast, if children are constructed as having contributions to make and rights to be fulfilled, then policies recognise children as individuals in their own right (and not solely dependents on their parents), for the integrity of their own persons, reputation and status, and for their own participation in decisions that affect them. Children's rights, and particularly the UNCRC, can be analysed for presenting their own

constructions of childhood, which present children as being rights-holders but entitled 'to special care and assistance'.

- More radical social movement ideas of oppression and discrimination have been slower to come to childhood studies but are beginning to be articulated by the concept of adultism. Essentially the concept recognises that embedded power relations create attitudes, systems and institutions that privilege adult norms; as Sundhall describes:
- Thus, adulthood becomes naturalized and at the same time age becomes a legitimate power order when it comes to the age categorization of children, since children's subordination is regarded as something natural and often desirable; children are viewed as 'under development' and in need of adult protection and care.
- Attitudes, systems and institutions need not be maliciously intended, and the people involved may be acting with the best of intentions. But the privileging of adult norms can result in systematic disadvantages, discrimination against or oppression of children. To give examples from the UK, 'mosquito devices' remain available to be bought online in the UK, for individuals or businesses who want to dissuade children from using certain spaces. The devices emit a noise irritating and potentially painful to children and young people, but which are generally not heard by adults. The school starting age for children remains low across UK jurisdictions (e.g. age five in Scotland), in comparison to other European countries. This is justified primarily as encouraging children's literacy and numeracy skills, but there are strong counter-arguments that suggest this is too early for many children to benefit, children who start school later than their peers do better academically, and play-based provision would be of greater benefit to children.Child poverty is in fact family poverty but children can be disadvantaged in social security systems that penalise parents who are not taking on paid employment or by the 'two-child benefit cap' in the UK that restricts the amount of social security benefit that larger families can receive.By concentrating on childhood as a social category, these problematic dimensions are illuminated.
- Children are generally not recognised as political agents, are excluded from democratic voting at younger ages and policy-making, and thus risk having their views and issues side-lined.Further, the concept of intersectionality shows how oppression by age and stage can intersect with other forms of oppression, which are more often rehearsed, such as gender, sexual orientation, ethnicity and race. Thus concepts of adultism are particularly challenging to current constructions of child–adult relations: certain accepted modes of policy and practice can be analysed as failing to meet or indeed breaching children's rights – even if the modes were developed expressly to improve children's lives.

- These ideas of adultism can be applied to the societal response to disasters. For example, reviews of climate change research found that children were positioned as passive victims with no substantive roles to help address climate risks.Based on their extensive experience in places that have faced disasters around the world, the international non-governmental organisation (NGO) PLAN powerfully writes:
- During a disaster, the physical survival needs of children (safe water, food, shelter, clothing, primary health care) are usually given a very high priority. But observation indicates that other needs and rights, which are also essential for children, like being protected from abuse and harm, education, rest, leisure, privacy and the right to participate freely in matters which affect them are too often overlooked.
- The analysis goes further, arguing that the disaster literature tends to confine children to three areas. First, they are subsumed under studies of women. Second, the literature provides a 'medicalised' narrative of disaster that limits children to social work and trauma responses, all focusing on how adults can return children to 'normal life' as soon as possible. Third, children become the media's representation of disasters – icons of passive and needy victims who need rescuing. Thus, children's physical survival may well be prioritised in disaster responses but not their wider range of human rights; they may be iconic in public perceptionsbut their perspectives and contributions under-acknowledged; and efforts are to return to the children to 'normal' childhood, rather than asking more fundamental questions of what childhood is and what it could and should be.
- Disasters can be highly testing to the regular practices of Government and services,particularly as the widescale and pervasive nature of the COVID-19 pandemic was unfamiliar in Scotland. The timescales for reacting and being proactive were far shorter than usual government business and democratic deliberation; the lack of sound research and other reliable information to base sound decisions was both unsettling and problematic. The infrastructure of government and service providers were themselves shaken as individuals were affected personally, adaptations needed to be swiftly made, and information and policy implementation had to be rolled out at great speed. Governance and government have their own complexities: for example, Scotland has devolved powers, with others reserved to the UK Government (such as on asylum and refugee policy); the Scottish Government and Parliament have national policy responsibility, but responsibility for implementation largely falls to other public bodies including local government. These very real-world constraints and complexities need to be recognised in considering the policy responses to COVID-19, for children in Scotland. No country has been identified, at the time

of writing, as being exemplary in holding up children's human rights during the COVID-19 pandemic.

- As Scotland moves forward with UNCRC incorporation into its domestic law, questions of State accountability to children and for children's rights become even more pressing. The power of the social science concepts and insights reviewed here is that they unveil underlying assumptions and suggest alternatives. As applied below, they provide ways to consider the independent CRIA's findings, in how the emergency policies constructed children and childhood and the ensuing impact on children's human rights – and to outline alternatives that position children and childhood in different ways that better respect, safeguard and facilitate their rights.

11.4 CHILDREN UNDER COVID-19 POLICIES:HIDDEN IN FAMILIES

- Lockdown restrictions resulted in the closure of public spaces and public institutions to children. Playgrounds were shut; children, as well as adults, were restricted to one bout of outdoor exercise a day; most public spaces for recreation and leisure, retail or arts were closed and, because of concerns to maintain social distancing, online grocery shopping was encouraged or, at most, one adult was permitted to shop alone. Further, institutions and spaces of state intervention, support and services were largely closed including: schools and early year provision (except for children deemed 'vulnerable' and/or had parents who worked as key workers); youth and community services; and certain court, tribunal, other formal review and appeals processes, and birth registrations. The State retreated and children were 'hidden' within their family households.
- Qvortrup wrote in 1994 about how children were familialised in the Global North, conceptually and morally placed within their families even though they were increasingly institutionalised in early years and school settings for most of their years and waking hours. Children's geographies have a long history of research that finds children and especially youth routinely excluded from public spaces populated by adults, whether commercial or in the open air, in the Global North.The COVID-19 policies went even further, as they closed public spaces and institutional places for children and confined children physically to their family households, making children increasingly dependent on their parents or other care-givers in their households. With this shift from public to private spaces for children, implementation of children's human rights depended even more on their households' living circumstances and their parents/ carers, with less direct reach from the State in terms of supporting services.

The good intentions of Government to respond to COVID-19 were firmly stated, which included considerations of children and (in Scotland) of taking a human rights approach. The Government went beyond PLAN's critique of disaster responses, as children's rights to be protected from abuse and harm were considered as well as their rights to physical survival. Nonetheless, children remained at risk of not having all their rights respected, recognised, and fulfilled, as policies so firmly constructed them as being best within their families and excluded from public spaces and institutions. This had ensuing repercussions for children who were at risk within their families or who were out of place. This suggests that children's human rights have not yet permeated how decision-makers respond, with responses returning to needs-based and wellbeing concerns with the emergency response and many other human rights of children being set aside.

- Early on, the Scottish Government recognised that this posed particular risks to certain children. Families are not safe places for all children, with particular concerns about child protection and domestic abuse. Both State social work services and NGO services were charged to support children and families in these situations: for example, through extending funding for helpline services available digitally and by telephone, to NGOs to provide further information and services, and hiring extra social workers.[33] While family households might generally be the best way to protect children, it was recognised that for some children the increased reliance on them made children more – rather than less – at risk. Thus, the Scottish Government went beyond most responses to disasters (see above), to consider children's rights to be protected from abuse or harm.
- The reliance on parental care and family households becomes even more evident, when considering how policies failed to meet the rights of children who were 'out of place'. One State response was not to recognise children *as* children if they were in the wrong places: initially, young people who were 16 or 17 years old were treated as adults and could receive Fixed Penalty Notices for breaking lockdown restrictions. Such notices could result in a fine and a criminal trajectory. After this was raised as a children's human rights issue, this provision was amended but not retrospectively for any children who had been found in breach. The State failed to respond to groups of children whose rights were already precarious. During the initial policy responses to the pandemic, there was little focus on the rights of children of refugee and asylum seekers or the rights of Gypsy/ Traveller children, even though research pre-pandemic identified these groups as being at particular risk of rights' violations, and that these risks were likely to be exacerbated by the COVID19- pandemic and the policy responses.[35] Policies often assumed 'traditional' ideas of families and of childhood. Reports of supermarkets challenging parents who shopped with children revealed assumptions about

family functioning and forms that do not hold true for all children.[36] There were assumptions that only adults had caring responsibilities, leaving young people with caring responsibilities struggling to access support for themselves and those they cared for.[37] Young people faced barriers to shopping online (which required a minimum shop for example and payment by debit or credit card) or difficulties in accessing shops in person. In other words, a strong construction of children as being safe in and dependent on their families, in their households, underpinned COVID19- policies. This exacerbated problems and undermined the rights of the many children whose circumstances did not fit into traditional constructions of childhood and families, in certain types of household spaces.

- The disaster literature suggests that children's rights to health tend to be prioritised in disaster responses but not their full range of rights. The independent CRIA analysis similarly found an extensive range of children's rights that policies had removed or ignored. What was surprising was that so many of them seemed *unnecessarily* removed or ignored. For example, the Lord President's guidance on children's contact with their non-residential parent made no mention of considering a child's best interests as the paramount consideration (the standard for court orders under S.11 of the Children (Scotland) Act 1995 for family law) or even as a primary consideration as required by Article 3 of the UNCRC. Birth registration was suspended, even though birth certification is known to contribute significantly to child protection and preventing child trafficking. Despite Scotland's lead in involving children and young people, individual rights to have views considered in decision-making were often deleted from requirements or not included in new ones, while the anger of young people (as well as others) at how school exams were cancelled and appeals handled eventually led to a Government commissioned review. The value of the CRIA process was demonstrated when new health guidance was changed, after intervention by the Commissioner, so that it became more rights compliant.[42] Thus, in the immediacy of COVID19- policy responses, a range of children's rights were not always addressed. Instead, they were undermined, and even deleted, despite being in place pre-pandemic. Other rights to provision – whether for education, mental health provision and reviews, access to play – and to participation – were often not substantially met in the policies. The lack of attention given to *all* of children's rights is troubling, especially as under the UNCRC, children's rights are recognised not only as interdependent but of equal importance.[43] The independent CRIA does not identify why this full range of children's rights were not considered, as the rationale was not documented in the accessed policy papers. As discussed above, the COVID19- pandemic created immense pressures on the State and disrupted the established ways of decision-making, planning and implementation in Scotland. But the analysis here points out that some of the gaps might have been avoidable if children's human rights had been given greater prominence and if a rights-based approach were embedded in systems, institutions, and ways of working.

- The independent CRIA did not have all the evidence that it needed, which was understandable but limiting. Evidence that had been collected prior to the pandemic might be indicative but not necessarily conclusive. Evidence collected in the early months of the pandemic was not robust by academic standards, either because it had been collected for policy-influence rather than research reasons or was in the early stages of collection and/or analysis. More problematically, the process highlighted how little of the available data available answered children's human rights questions nor was the data sufficiently disaggregated to know whether or how particular groups were at risk of having rights' violations. In Scotland and the UK more generally, there is considerable investment in statistical data collection both in administrative data and funded large-scale longitudinal and cohort studies. But these have not been constructed with children's human rights in mind. Children are excluded from many regular large-scale surveys as respondents (e.g. Scottish Crime Survey). Scotland does not have robust systems for collecting data on how State duty-holders respect, protect and fulfil children's human rights. Such realisations underline the need to reconsider statistical investments and design a suite of children's human rights indicators for Scotland, which can allow for constructive monitoring and accountability, once the UNCRC (Incorporation Scotland) Bill is implemented.

11.5 NOTION OF VULNERABILITY OF CHILDREN RIGHTS

- The notion of vulnerability is being reclaimed in legal philosophy, as a new and improved way to consider law.It is grounded in the observation that all people are vulnerable, that vulnerability is a universal and inevitable condition of being human, because we are embodied. If we recognise that all people are vulnerable and dependent, then law should be based on that premise rather than on the rational, autonomous and independent man assumed by traditional liberal theory of rights. According to Fineman, the State then must take on a more proactive role, than in liberal theory, to ensure care is valued, social relations encouraged, protection is given as needed and resources are provided to ensure resilience. Fineman contends that vulnerability offers routes to stronger obligations on the State to ensure equality.
- However, recognising vulnerability under COVID-19 policies did not always support the recognition or the realisation of children's human rights. As the independent CRIA found, the actions of the State produced and had disproportionate and negative consequences for children and even more so for particular groups of children, demonstrating a form of adultism. While

vulnerability may be universal, it was more acute for children as a group, and even more acute for particular groups of children.The independent CRIA's findings illuminated which children and which rights were made vulnerable as a result of the State's response to the pandemic. For example, restrictions in young offenders' institutions left young people confined to their cells, with very limited or no contact with family members, and access to education and other activities halted. Yet, the particular vulnerabilities of young people to suicide and mental health problems were not specifically considered nor their particular claim for early release as children.When the particularities of children's position and status were not considered, COVID-19 policies did not always ensure that the human rights of all children were equally protected.

- Vulnerability may well be an intrinsic part of the human condition; however, being labelled as vulnerable within current social structures does not generally result in children's empowerment nor does it necessarily advance their interests. Rather, the label of vulnerability often has negative consequences for children. It is frequently bound up with paternalistic discussions about children's capacity and is used as a rationale by adults to gatekeep and limit children's human rights, especially their participation rights.In social policy discourses, to be vulnerable remains the exception not the norm and rarely offers emancipation. There is stigma associated with being viewed or perceived as vulnerable. Vulnerability is used euphemistically to describe particular populations of children who are 'at risk' or who are 'on the margins'. During the pandemic, there were strong policy for and media discourses on 'vulnerable children'.Children categorised as vulnerable were permitted to attend school and early years provision. In this context, attendance was largely a safeguarding measure, prioritising children's protection and safety, rather than a means to realise children's rights to education, social relationships or other rights like play. However, far fewer children attended early years and school provision than had been anticipated, with indicative evidence of families not recognising themselves as vulnerable, not accessing information on the support available, not wanting to be labelled as 'vulnerable' and/or experiencing practical barriers like transportation to and from hub schools.Provision for 'vulnerable children' was not accessed as the State had anticipated.
- Fineman argues that inequalities are far from inevitable: rather society and its institutions produce and reproduce them. Thus, in its actions, the State has the potential to address and reverse inequalities. However, as discussed above, under COVID-19 policies the State largely retreated from children's lives and confined children to their families. Fineman argues families alone cannot mitigate or ameliorate individuals' vulnerability. As a structure, families are themselves vulnerable and, as the independent CRIA confirmed, families are not necessarily 'safe havens' for children. So, in retreating, rather than addressing inequalities, the

State produced and exacerbated certain inequalities experienced by children. In not meeting the positive obligations to children that the vulnerability thesis demands, the State created a 'vacuum' where children's human rights were increasingly exposed to breaches. This serves as a persuasive example of how the State, albeit perhaps inadvertently, responded in adultist ways that systematically disadvantaged children as a group.

11.6 AT A GLANCE

- Despite seemingly progressive advancement towards 'making children's rights real' in Scotland, the pandemic uncovered that fragility of embedding children's human rights in policy-making. The pandemic placed extreme pressures on the State, upturning normal processes of decision-making, implementation, and provision, particularly in a State based on democratic deliberations, governed by the rule of law, and having complex welfare provision. The distribution of powers between the UK Government, the Scottish Government and other public bodies added complexities in policy-making let alone for implementation. The scale of response needed was unprecedented in living memory in Scotland, and those involved in responding faced very real constraints and dilemmas. In this disaster situation, efforts to safeguard rights to survival and development were foremost. The rights of children to protection from abuse and harm were recognised. The resulting policies, however, led to other rights of children being disregarded and children as a group being disadvantaged.
- The concept of adultism recognises that children can be systematically disadvantaged not because of adults' bad intentions but because of the pervasiveness of adult norms and the resulting power relations. In the policy responses to the COVID-19 pandemic, children were not always recognised as rights-holders and thus their rights were not always supported and promoted. This was evident from the lack of participation of young people to decisions about exams and their own rights to appeal, to the closure of schools, to the remaining access to public and consumer services skewed towards adults. The lens of children's human rights makes explicit the risks to children, and particular groups of children, from COVID-19 policies. These policies largely constructed children and childhood as vulnerable, dependent and best protected in their families. This had significant and negative consequences for the human rights of children who were 'out of place' or whose childhoods did not fit with these normative ideas about childhood.
- Through its retrenchment, the State largely left children to be subsumed and made invisible in families. This recalibration left children and their human rights at risk. In making children increasingly dependent on their families, existing inequalities were exacerbated. The prioritisation of rights to survival and

development, understandable under disaster conditions caused by the pandemic, led to a widespread failure to recognise the equal value and interdependence of children's human rights. The independent CRIA revealed how rights to participation, education, birth registration, access to justice and to services were *unnecessarily* set aside. Children were excluded from decisions that had profound impacts on their human rights. These decisions and those who made them were, and remain, largely unaccountable to children. Taken together, this underscores the continued importance of recognising children and childhood as critical categories for policy and legal analysis, as well as recognising children as a social group with particular rights and needs that need particular safeguarding.

- Fineman's work on vulnerability promises routes to achieving substantive equality. Yet, our analysis found theorisations of vulnerability inadequate to tackle adultism that arises from largely adult-orientated systems and public institutions. Conceptualisations of 'vulnerable children' did not ensure that children's human rights were advanced. Ideas of universal vulnerability mask the unequal position that children have in society and in public institutions. Failure to acknowledge or attend to such structural issues risks further entrenching inequalities and the systematic disadvantages experienced by children and by specific groups of children. The independent CRIA found labels of vulnerability were sometimes rejected by children and their families because such labels continued to bring stigma, while other children and their families did not recognise themselves as vulnerable even though they might have been eligible. Rather than a lever for advancing the interests or position of children, we found vulnerability an oppressive categorisation, which may act as a further barrier to children claiming their human rights. Only perhaps when vulnerability is truly claimed as universal, by those currently in power and powerful, will it be useful to apply to childhood policy.
- A State that is positive and proactive in its obligations to its citizens is a cornerstone of Fineman's vulnerability thesis. However, the pervasive construction of childhood vulnerability in COVID-19 policies was not found sufficient to ensure the State did proactively meet its obligations to children. Accountability is necessary to a human rights framework, and indeed multiple forms of accountability, as part of the relationship between rights-holders and duty-bearers. At the core of a human rights framework is a recognition of human dignity and a requirement that, at a minimum, rights must be respected. Thus a human rights frameworks assists constructing childhood as worthy of attention and children's rights as requiring respect and protection.

- Due to emerging concerns in Scotland that COVID-19 policies did not recognise children's human rights adequately, the resulting independent CRIA assisted in making visible the systematic disadvantages experienced by children. In doing so, the independent CRIA made explicit the actual or potential breaches to children's human rights – and suggested ways to mitigate them. It provided learning for subsequent lockdowns and ensuing priorities. For example, recommendations were made on how to address digital exclusion, to prioritise outdoor learning and provision once restrictions were lifted, and to address food poverty. It illuminated gaps in children's human rights accountability, such as the lack of children's rights indicators and disaggregated data, accessible opportunities for children to appeal and complain when their rights were being breached, and information for and participation of children and young people. As Scotland moves to incorporating the UNCRC in domestic law, the independent CRIA shows the value of such impact assessments to surface, discuss and mitigate children's human rights violations and – even more importantly – maximise the fulfilment of their rights.

BIBLIOGRAPHY

- 1976, Aristotle: Metaphysics Books M and N, Oxford: Clarendon Press.
- 1979, Articles on Aristotle, Vol 3. Metaphysics, London: Duckworth.
- 1983, "Prime Matter: a Rejoinder," Phronesis, 28: 197–211.
- 1984, "The Aporematic Approach to Primary Being in Metaphysics Z," Canadian Journal of Philosophy, Suppl. Vol. 10: 1–20.
- 1985, "On the Origins of Some Aristotelian Theses About Predication," in Bogen and McGuire 1985, pp. 101–131.
- 1986, "Aristotle: Essence and Accident," in Philosophical Grounds of Rationality: Intentions, Categories, Ends, R. Grandy and R. Warner (eds.), Oxford: Clarendon Press, pp. 411–439.
- 1987, "Metaphysics and Logic," in Matthen 1987a, pp. 127–149.
- 1994, "Aristotle on Identity," in Scaltsas, Charles, and Gill 1994, pp. 41–54.
- 1994, "Matter and Form: Unity, Persistence, and Identity," in Scaltsas, Charles, and Gill 1994, pp. 75–105.
- 1995, "Potentiality in Aristotle's Science and Metaphysics," Pacific Philosophical Quarterly, 76: 405–418.
- 1995, "Science and the Science of Substance in Aristotle's Metaphysics Z," Pacific Philosophical Quarterly, 76: 419–469.
- 2001, A Map of Metaphysics Zeta, Pittsburgh, PA: Mathesis Publications.
- 2002, Aristotle on Meaning and Essence, Oxford: Clarendon Press.
- 2007, "Aristotle's Abstract Ontology," Newsletters for the Society for Ancient Greek Philosophy, 8 (2): 12–23.
- 2008, Aristotle's Divine Intellect, Milwaukee: Marquette University Press.

- 2011, "Why Aristotle's God is Not the Unmoved Mover," Oxford Studies in Ancient Philosophy 40, 91–109.
- 2018, "Activity, actuality, and analogy: Comments on Aryeh Kosman, The Activity of Being: An Essay on Aristotle's Ontology," European Journal of Philosophy, 26: 872–880.
- 2018, "Comments on Aryeh Kosman's The Activity of Being: An Essay on Aristotle's Ontology," European Journal of Philosophy, 26: 860–871.
- Ackrill, J. L., 1963, Aristotle: Categories and De Interpretatione, Oxford: Clarendon Press.
- Addis, L., 1972, "Aristotle and the Independence of Substances," Philosophy and Phenomenological Research, 54: 699–708.
- Adler, Joseph A., 1999, "Zhou Dunyi: The Metaphysics and Practice of Sagehood," in Sources of Chinese Tradition, William Theodore De Bary and Irene Bloom (eds.), 2nd edition, 2 volumes, New York: Columbia University Press.
- Ainsworth, Thomas, 2018, "Priority in Being in Aristotle," Philosophy Compass, 13: 1–10.
- Albritton, Rogers, 1957, "Forms of Particular Substances in Aristotle's Metaphysics," Journal of Philosophy, 54: 699–707.
- Alexandrou, S., 2014, Aristotle's Metaphysics Lambda: Annotated Critical Edition, Leiden: Brill.
- Allen, R. E., 1969, "Individual Properties in Aristotle's Categories," Phronesis, 14: 31–39.
- Ames, Roger T. and David L. Hall (trans.), 2003, Dao De Jing—Making This Life Significant—A Philosophical Translation, New York: Ballantine Books.
- Ames, Roger T., 2011, Confucian Role Ethics: A Vocabulary, Hong Kong and Honolulu: Chinese University of Hong Kong Press and University of Hawai'i Press.
- Anagnostopoulos, Andreas, 2011, "Senses of 'Dunamis' and the Structure of Aristotle's 'Metaphysics' Theta," Phronesis, 56: 388–425.
- Anagnostopoulos, Georgios (ed.), 2009, A Companion to Aristotle, Chichester: Wiley-Blackwell.
- Angioni, Lucas, 2014, "Definition and Essence in Aristotle's 'Metaphysics' vii 4," Ancient Philosophy, 34: 75–100.
- Angle, Stephen C. and Justin Tiwald, 2017, Neo-Confucianism: A Philosophical Introduction, Cambridge: Polity.

- Angle, Stephen C., 2009, Sagehood: The Contemporary Significance of Neo-Confucian Philosophy, Oxford: Oxford University Press.
- Annas, J., 1974, "Individuals in Aristotle's Categories: Two Queries," Phronesis, 19: 146–152.
- Anscombe, G. E. M., 1953, "The Principle of Individuation," Proceedings of the Aristotelian Society, Supplementary Volume 27: 83–96; reprinted in Barnes, Schofield, and Sorabji 1979, pp. 88–95.
- Aranyosi, István A., 2004, "Aristotelian Nonsubstantial Particulars," Philosophical Writings, 26: 3–12.
- Bäck, Allan, 2004, "What is being QUA being?," Poznan Studies in the Philosophy of the Sciences and the Humanities, 82: 37–58.
- Baghdassarian, F., 2019, Aristote Metaphysique Lambda, Paris: Vrin.
- Bai, Tongdong, 2008, "An Ontological Interpretation of you (Something and wu (Nothing) in the Laozi," Journal of Chinese Philosophy, 35(2): 339–51.
- Bambrough, R. (ed.), 1965, New Essays on Plato and Aristotle, London: Routledge and Kegan Paul.
- Banka, Rafal, 2011, "Ontological Intuitions and Their Interpretational Problems: The Case of dao," in Inter-Culturality and Philosophic Discourse, Yolaine Escande, Chenyang Li, and Vincent Shen (eds.), Cambridge: Cambridge Scholars, pp. 81–95.
- Barnes J., M. Schofield, and R. R. K. Sorabji (eds.), 1975, Articles on Aristotle (Volume 1: Science), London: Duckworth.
- Beere, Jonathan, 2009, Doing and Being: An Interpretation of Aristotle's Metaphysics Theta, Oxford: Oxford University Press.
- Behuniak, James, 2004, Mencius on Becoming Human, Albany: State University of New York Press.
- Berti, E., 2001, "Multiplicity and Unity of Being in Aristotle," Proceedings of the Aristotelian Society, 101: 185–207.
- Billioud, Sébastien, 2012, Thinking Through Confucian Modernity: A Study of Mou Zongsan's Metaphysics, Leiden: Brill.
- Block, I., 1978, "Substance in Aristotle," in Paideia: Special Aristotle Issue, G. C. Simmons (ed.), Brockport, NY, pp. 59–64.
- Bodéüs, R., 2000, Aristotle and the Theology of Living Immortals, Albany: SUNY Press.
- Bogen, J. and J. E. McGuire (eds.), 1985, How Things Are: Studies in Predication and the History of Philosophy, Dordrecht: Reidel.

- Bolton, R., 1994, "Aristotle's Conception of Metaphysics as a Science," in Scaltsas, Charles, and Gill 1994, pp. 321–354.
- Bordt, M., 2006, Aristoteles Metaphysik XII, Darmstadt: WBG.
- Bostock, D., 1994, Aristotle: Metaphysics Books Z and H, Oxford: Clarendon Press.
- Bowin, John, 2008, "Aristotle on Identity and Persistence," Apeiron, 41 (1): 63–90.
- Brindley, Erica Fox, 2012, Music, Cosmology, and the Politics of Harmony in Early China, Albany: State University of New York Press.
- Broadie, S., 1993, "Que fait le premier moteur d'Aristote?" Revue philosophique de la France et de l'étranger 183: 375–411.
- Brody, B. A., 1973, "Why Settle for Anything Less Than Good Old-fashioned Aristotelian Essentialism?" Noûs, 7: 351–365.
- Brook, Angus, 2015, "Substance and the Primary Sense of Being in Aristotle," Review of Metaphysics, 68: 521–544.
- Buchheim, T., 2001, "The Functions of the Concept of Physis in Aristotle's Metaphysics," Oxford Studies in Ancient Philosophy, 20: 201–234.
- Burik, Steven, 2018, "Logos and Dao Revisited: A Non-Metaphysical Interpretation," Philosophy East and West, 68(1): 23–41.
- Burnyeat, M. F. et al., 1979, Notes on Book Zeta of Aristotle's Metaphysics, Oxford: Sub-faculty of Philosophy.
- Cameron, Margaret, 2019, "The Moving Cause of Artefacts: The Role of 'Techne' in Metaphysical Explanation," Ancient PhilosophyToday, 1: 64–83.
- Chai, David, 2010, "Meontology in Early Xuanxue Thought," Journal of Chinese Philosophy, 37(1): 91–102.
- Chappell, V., 1973, "Aristotle's Conception of Matter," Journal of Philosophy, 70: 679–696.
- Charles, David, 1993, "Aristotle on Substance, Essence, and Biological Kinds," Proceedings of the Boston Area Colloquium in Ancient Philosophy, 7: 227–261.
- Charlton, W., 1972, "Aristotle and the Principle of Individuation," Phronesis, 17: 239–249.
- Chen, Chung-Hwan, 1957, "Aristotle's Concept of Primary Substance in Books Z and H of the Metaphysics," Phronesis, 2: 46–59.
- Code, Alan, 1978, "No Universal is a Substance: an Interpretation of Metaphysics Z 13, 1038b 8–15," in G. C. Simmons 1978, pp. 65–74.

INDEX

J

M

N

O

P

R

S

T

U

V

W

www.ingramcontent.com/pod-product-compliance
Lightning Source LLC
LaVergne TN
LVHW062121200926
848484LV00007B/1735